The Case of Galileo and the Church

WALTER CARDINAL BRANDMÜLLER

THE CASE OF GALILEO AND THE CHURCH

Translated by Michael J. Miller

SOPHIA INSTITUTE PRESS
Manchester, New Hampshire

First published in German as *Der Fall Galilei und die Kirche*
Illertissen, Germany: Media Maria Verlag, 2021.

Cover by Updatefordesign Studio
Cover image: AI Generated with Midjourney — by the designer (Flavius Petrisor)

Sophia Institute Press
Box 5284, Manchester, NH 03108
1-800-888-9344
www.SophiaInstitute.com

Sophia Institute Press is a registered trademark of Sophia Institute.

paperback ISBN 979-8-88911-312-6

ebook ISBN 979-8-88911-313-3

Library of Congress Control Number: 2024945068

First printing

CONTENTS

Foreword

Probably the observation that Bertolt Brecht's *Life of Galileo*, which was originally staged many years ago now, is still performed again and again was what prompted Media Maria Verlag to suggest a new edition of my presentation of that cause célèbre, published in 1994. In fact Brecht's famous play has a place as much now as ever in theater programs. For example, it had a premiere — with music by Hanns Eisler — in Düsseldorf on January 16, 2020. Frank Castorf staged a new production in Berlin, and Brecht's *Galileo* was performed again recently in Cologne, Essen, and Saarbrücken. Ingo Langner just took up the theme in his play *Mikado*. Evidently the problems of "science and faith," "progress and the Church," "ethics and technology" occupy inquiring minds as much now as ever.

Yet if someone wanted to take Brecht as guide on this tour of an intellectual mountain range, it would be good to know why, when, and for what purpose Brecht wrote his *Galileo*. He was not concerned about the astronomer and the Inquisition — he was concerned about himself. As Ingo Langner explains in his foreword to *Mikado*,

> In Autumn of 1938 Bertolt Brecht was accused of speaking up publicly against Hitler's brown-shirt terror but remaining silent about Stalin's Red terror. Brecht reacted to this with the drama *Life of Galileo*.

> Only apparently does it deal with the rebellion of reason against the dogmatism of the Roman Catholic Church. Actually Brecht intends thereby to justify his own silence about Stalin's show trials in Moscow. Concealed in the figure of the polymath Galileo Galilei is the playwright Bertolt Brecht.

In fact, someone like Bertolt Brecht is capable of contributing no more to our understanding of the historical Galileo case than Friedrich Schiller can with his *Don Carlos* to our understanding of the unfortunate son of Philip II of Spain. Yet it is a fact that, again and again, the "educated middle class" owes its image of Galileo, including his religion, not to a study of the historical record but rather ... to the theater!

This is precisely why a scientific, dispassionate presentation of the famous case should be contrasted with the Galileo of drama and fiction.

In this case, too, historical truth has found it particularly difficult to win respect for itself, since "Galileo" is conjured up again and again when, out of ideological prejudice, arguments against the obscurantist Church are to be made from this case.

Considering all this, of course, it should be noted that meanwhile, precisely among proponents of the theory of science, an understanding of the famous case has prevailed which carefully distinguishes the highlights and shadows. For example they acknowledge that the cardinals of the Inquisition were in error with their literal interpretation of Sacred Scripture concerning a fixed earth, while Galileo, the astronomer, could not prove what he maintained. Galileo, furthermore, was right with his pertinent interpretation of the biblical texts. The Inquisition, in contrast, ultimately had the broader view when it demanded of Galileo only that he should not champion his opinion as certain knowledge, but rather should present it as a hypothesis for debate. For even supposing that Galileo's arguments in fact agreed with the cosmic reality, that would by no means rule out the possibility that another explanation might be even more convincing.

Thus the Inquisition, with its conviction that as a matter of principle scientific knowledge can become outdated, proved to be the spearhead of the intellectual vanguard.

> Since the first publication of this book in 1994, twenty-five years have now passed, in which the topic "Galileo" has lost none of its interest. Moreover a series of pertinent studies has appeared which, generally speaking, treat individual aspects of the topic. These are for the most part astronomical and physical aspects. Many also have to do with the question that we are posing. In our case, though, we are concerned with the rather narrowly framed question of the relation of Galileo's person and work to the ecclesiastical authorities.

The present new edition contains unaltered the text from 1994. A bibliographical supplement has been added, for which I am greatly indebted to Professor Dr. Johannes Grohe. May this new edition, in which the typographical errors of the first edition have been corrected, stimulate readers to deal with this topic anew.

Walter Brandmüller
Rome, January 2021

The Case of Galileo and the Church

INTRODUCTION

IN 1992, AT the invitation of the Libreria Editrice Vaticana, I submitted an Italian translation of my book *Galilei und die Kirche oder Das Recht auf Irrtum* (Galileo and the Church, or the Right to Error), which had been published in German in 1982 by Verlag F. Pustet in Regensburg. A Spanish translation of the same book under the title *Galileo y la Iglesia* had been published in 1987 by Rialp, in Madrid.

The 1992 version with the title *Galilei e la Chiesa—ossia il diritto ad errare* was essentially different from these two editions. Not that the view of the subject that I had originally presented or my judgments on the persons and events of that time had to be corrected. Rather, there were two practical reasons. First, I had to sort out the literature on the topic that had appeared since 1982 because of the colloquia, symposia, etc. that had been organized in 1983 to commemorate the trial of Galileo in the year 1633, and to adopt their findings. Their great value consists above all in the fact that they not infrequently contribute important and hitherto unknown information, mostly for the scientific or intellectual- and cultural-historical evaluation of the figure of Galileo. Secondly, though, the book that had appeared in 1982 was expanded with new chapters presenting the historical aftermath and the definitive decision on the Galileo case that was finally issued in 1820.

Shortly thereafter, at the instigation of the Pontifical Academy of Sciences, the publisher Olschki in Florence was able to print the complete edition of the dossier that had been produced over the course of the discussions about the work *Elementi di Ottica e di Astronomia* by the German-Roman astronomer Giuseppe Settele that had been conducted within the Roman Curia, particularly within the Holy Office: Wilhelm Brandmüller and Egon Johannes Greipl, *Copernico, Galilei e la Chiesa: Fine della controversia (1820): Gli Atti del Sant' Uffizio* (Florence: Olschki, 1992).

Since both books had appeared only in Italian, the next step was obviously to make them and—as far as the edition was concerned—their findings available to the German-speaking public also. This is done now in such a way that the presentation of the Galileo case until the judgment in 1633 was taken essentially from the first-mentioned book, while the aftermath down to the resolution of the question in the year 1820 corresponds, with unimportant exceptions, to the commentary that I had composed for the edition of the *Acts*.

The fact that the Holy Office decided the question, "Is the heliocentric system of the cosmos compatible with the Catholic faith?" once and for all in the affirmative is almost never mentioned in the pertinent literature and is not appreciated at all, even though it has been well known for a long time.[1] The contemporary press had already reported it more or less correctly.[2] Travel literature and memoirs mentioned the event here and there,[3] and in a "Speech about the Boundary of the Sciences," held on November 28, 1855, during the Forum of the Bavarian Academy of Sciences, Friedrich von Thiersch dealt also with

1 See Antonio Favaro, ed., *Le Opere di G. Galilei, Edizione Nazionale*, vols. 1–20 (Florence, 1890–1909) at 19:419–421. Volume 19 contains the official records of the trial.

2 See ibid., 238–239.

3 See, for example, Wilhelm Christian Müller, *Briefe an deutsche Freunde von einer Reise durch Italian … 1820 und 1821*, II (Altona: J. F. Hammerich, 1824), 697, 714–716 (incorrect and arrogantly "enlightened" in its tenor).

the Galileo-Settele matter.[4] Thiersch had been staying in Rome during these events, and Reinhold Niebuhr and other acquaintances were his sources of information in this regard. His talk in Munich is not correct on all points, but its general drift shows understanding and respect for the curial procedures when he writes: "As far as we know, this decision has been accepted without contradiction in all countries of Catholic Christendom, and it has rightly been asserted as a proof that the Catholic Church, too, does not shut itself off from scientific discoveries and avoids taking a stance that contradicts the acknowledged results of prudent, well-founded research."[5]

Then Maurizio Benedetto Olivieri, who as a Commissioner of the Holy Office personally had played a leading role in this decision, later reported on it and gave reasons for it in a shorter work that was published posthumously in 1872.[6] In doing so he was able to rely on archival material of the Holy Office, since he had been a commissioner of that dicastery for decades. Franz H. Reusch was able to rely on Olivieri's and Thiersch's remarks when he set down in writing his studies about the Galileo trial and about the Index.[7]

4 See Friedrich von Thiersch, "Rede über die Grenzscheide der Wissenschaften," *Gelehrte Anzeigen der Königlich Bayerischen Akademie der Wissenschaften* 41 (1855): 179, 189–191.

5 Ibid., 191.

6 See *Di Copernico e di Galileo, Scritto postumo del P. M. B. Olivieri, Ex-Generale dei Domenicani e Commissario della S. Romana ed Universale Inquisizione, ora per la prima volta messo in luce sull'autografo per la cura di un religioso dello stesso istituto* (Bologna: G. Romagnoli, 1872). An inaccurate reference to the failure to distribute the work is found on the pertinent card in the catalogue of the Vatican Library, catalogue number: Ferraioli IV 4897; the name of the publisher Fr. Tommaso Bonora, O.P., is found in the catalogue of the Biblioteca Casanatense.

7 See Franz H. Reusch, *Der Process Galilei's und die Jesuiten* (Bonn: E. Weber, 1879), 441–442; Franz H. Reusch, *Der Index der verbotenen Bücher: Ein Beitrag zur Kirchen- und Literaturgeschichte* II/1 (Bonn: Max Cohen & Sohn, 1885), 400.

A further source for our knowledge of these events, which substantially completes the "arid" documents, was made available in part in 1886 by the important Roman philologist and literary historian Giuseppe Cugnoni.[8] We are talking about the voluminous diary of the man who got the stone rolling in 1820: the ethnic-German Roman and professor of astronomy at the "Sapienza," Giuseppe Settele.[9] By his "fight" for an imprimatur for his book *Elementi di Ottica e di Astronomia,* which in volume 2 advocated the Copernican system of the cosmos, he categorically demanded the aforementioned decision. In his diary he gives a thorough description of the procedures that led to the granting of the imprimatur for his book. Cugnoni published several passages from this diary, and in doing so sketched a brief biography of Settele.

A little later Antonio Favaro, whom Cugnoni had put on the right track, also dealt with Settele and his diary in the course of his own Galileo research and published his findings under the very tendentious title, *L'ultima fase della lotta contro il sistema copernicano* (The last phase of the fight against the Copernican system), which distorts the subject matter, or even turns it upside down.[10] The corresponding "Deliberazione della Congregazione del S. Uffizio intorno alla stampa dei libri che insegnano il moto della terra" (Deliberations of the Congregation of the Holy Office concerning the printing of books that teach the earth's movement) from the years 1820–1822 were edited and published by Favaro in volume 19 of his *Edizione Nazionale* of the works of Galileo Galilei.[11]

8 See Giuseppe Cugnoni, "Giuseppe Settele e il suo diario," *La Scuola Romana* 12 (1886): 265–284. On Giuseppe Cugnoni (1824–1908), see Alessandra Cimmino, in *DBI* 31 (1985): 338–342.

9 See ibid., 221–223. The "Sapienza" (Italian for "wisdom") is the common term for the venerable University of Rome.

10 Published in *Memorie del Reale Istituto Veneto di Scienze, Lettere ed Arti* 24 (1891): 419–430.

11 See note 1 above.

Now, after Settele's diary had passed through several hands, it ended up in the possession of the University of Rome, where it was restored in 1977–1978 and preserved.[12] This circumstance enabled Jole Vernacchia-Galli in 1984 to use the diary as a source for her history of the Sapienza in the years 1810–1836.[13] Of course she also discusses in it the affair in 1820. However, unfortunately, she omitted a commentary on the secondary literature. The same is true also of the faulty reprinting of the relevant parts of the diary by Paolo Maffei in 1987.[14] Nonetheless, this source for the events that led to the momentous decision in 1820 thus became available for the first time.

Both authors reprinted also in their works a photocopy of the dossier compiled by the Commissioner of the Holy Office, Fr. Olivieri,

12 See Jole Vernacchia-Galli, *L'Archiginnasio Romano secondo il diario del prof. Giuseppe Settele (1810–1836)*, Studi e Fonti per la storia dell'Università di Roma 2 (Rome: Edizioni dell'Ateneno, 1984), 9–10; Paolo Maffei, *Giuseppe Settele, il suo diario e la questione Galileiana* (Foligno: Edizioni dell'Arquata, 1987), 47–73.

13 See note 12.

14 See note 12. It is highly regrettable and downright detrimental that the cause célébre is taken up also by authors who lack the necessary scholarly tools. This is especially true about the work by Paolo Scandaletti, *Galileo privato* (Milan: Camunia, 1989), in which the chapter "La vittoria postuma" is particularly objectionable. One example suffices. On page 238 alone the reader finds the following crude errors: (1) Scandaletti claims that, at the urging of Cardinal Leopoldo de' Medici, Benedict XIV ordered that the prohibition of heliocentric books be stricken from the Index. This happened in 1757, whereas Leopoldo de' Medici had already died in the year 1675! (2) Cardinal Galli's reply to Lalande in 1765 speaks about a decree of the Holy Office and not, as Scandaletti writes, about the decree of the Holy Office in 1616 that condemned heliocentrism as absurd and heretical. Such a decree does not exist! (3) Scandaletti claims that Settele requested the imprimatur because he was a priest. This was necessary for books by all authors, not only for those of clerics. (4) Settele's book does not begin with the words "*Movendosi la terra*," as Scandaletti writes, but rather the section that deals with cosmology begins that way. Similar inaccuracies *on one and the same page*, which the present author happened to notice by chance, are quite characteristic of this book.

which served the members of the Holy Office as a basis for their deliberations and decision. Incidentally, several copies of this dossier had been printed for official use, and besides the one in the Archive of the Holy Office, others were still accessible in the Biblioteca Apostolica Vaticana and in the Biblioteca Universitaria in Perugia, although they had never been used; it goes without saying that this dossier did not contain all the material that accrued over the course of the proceedings. This material, to the extent that it has been preserved, is now available in its entirety to interested researchers in the aforementioned edition.

While working on these projects we were able to enjoy the sympathetic cooperation of the Congregation for the Doctrine of the Faith, the successor to the Holy Office and the Congregation of the Index. For this our most courteous thanks go to its then-prefect, His Eminence Joseph Cardinal Ratzinger.

Naturally we are grateful to the Archivist of the Congregation, Fr. Innocenzo Mariani, and to Msgr. Helmut Moll, who discovered the needed archival material with tireless patience and helpfully presented it for our use. Thus it can be assumed that completeness was achieved in the process.

Nevertheless the critical reader will get the impression here and there that more archival material must have been available than was cited here. Of course we should take into account, first, that the archive suffered considerable losses from the time before the Congress of Vienna when it was confiscated and taken to Paris, then shipped back to Rome and moved repeatedly within Rome. On the other hand, we should point out that the officials and cardinals involved in the proceedings usually kept their documents at home, which is why it was probably rare for them to end up in the archive of the Congregation. Finally, the business was also conducted to a great extent orally and not according to the official practice that later became conventional. For all these reasons there are no doubt fewer written records of the proceedings to be discussed here than one might wish.

It would have been highly desirable also to be able to consult the archives of the Accademia dell'Arcadia and of the Odescalchi family. One suspects that both contain material that could illustrate the proceedings in the years 1820–1822. In one case — the Accademia dell'Arcadia — the organization of the archive has not yet progressed far enough to make use of it possible, and in the other case I received no answer to two inquiries.

I owe a debt of gratitude, on the other hand, to the General Archives of the Dominicans at Santa Sabina, where I was able to find some bits of information about Olivieri, but particularly his portrait.

Part One

The Case

Chapter 1
The Trauma

The name *Galileo* has become a symbol. Whenever he is mentioned, it conjures up on the horizon the conflicts that his case precipitated between natural science (or science in general) and freedom of inquiry and of thought, on the one hand, and faith and the Church, on the other hand. Here Galileo stands for freedom, progress, modernity, while the Church stands for rigid dogma, shackled thought, and obscurantism. Indeed, some even speak about bending the law, falsifying documents, and brainwashing. A sober, dispassionate discussion about the Galileo case often seems quite impossible nowadays.

The emotions that almost always surge when this name is invoked, however, hint at how deeply the event involving Galileo touches the self-understanding of modern man; we can probably speak about psychological, intellectual trauma — both on the part of those who side with science, and also on the part of theologians and men of the Church.

First, it was probably the natural scientists who felt that they were condemned in the person of Galileo and saw their science, their intellectual life, called into question and discredited. Citing the Galileo case, however, was also an all-too-convenient moral justification for breaking publicly with the Church or else silently leaving it, which often had occurred long since for other reasons.

However it should be noted even here that such developments were not an immediate consequence of the Galileo judgment. Rather, it can be shown that natural scientists turned away from faith and Church in a process that started only as a result of the Enlightenment and reached its high point around the turn of the twentieth century. Buoyed up by the breathtaking achievements of physics, chemistry, biology, and medicine which the scientific-technological-industrial nineteenth century experienced, a mentality developed in what were then called the "educated professions" [*"gebildeten Ständen"*] that was inclined to explain now not only the world that can be counted, measured, and weighed, but all of reality, with the methods of science and to measure it by those standards. This pulled the rug out from under religion and challenged it, insofar as it relied on divine revelation.

Since this intellectual attitude went hand in hand with what were perceived as spectacular advances in the natural sciences, it soon became a worldview; a so-called scientific cosmology was formed, and without further inquiry the laws of nature became the first and highest principles. And this was true until after World War II; the end of the war, in conjunction with the reconstruction of ruined Europe that had become necessary, introduced a new phase of scientific discoveries and technological inventions and thus refueled the belief in the scientific, technological, and economic feasibility of all things.

Of course, the shock of Hiroshima and Nagasaki had given rise to initial doubts.

These preconditions have defined to this day the position taken up by very broad, scientifically educated circles with regard to the Galileo case, although ecological problems are increasingly acknowledged, or are viewed with outright anxiety.

For a long time numerous statements by scientists on this matter, by their aggressive tone and their utter unwillingness to become objectively informed about the Catholic faith, clearly manifested the profound effect of the Galileo trauma. Even leading scholars, when coming to terms with faith and the Church, dispensed with the most elementary requirements for an intellectually honest dialogue—a method that they would never follow, much less approve of, within their own field of specialization. The physicist Walter Gerlach, for example, wrote about the Copernican system that it offended "especially against two" dogmas: "The first was the division of the cosmos into the sublunary sphere, the imperfect and sinful world on earth, and the perfect, ethereal spheres which remained eternally the same.... The second dogma was the stationary earth and the sun moving around it—unmistakable theses in the Bible and Sacred Scriptures."[15]

However, at no time and in no context did the Church ever formulate such dogmas; at most these were features of ancient and medieval cosmology that Galileo's contemporaries typically held. Of course it would be necessary first to obtain information about the concept of dogma before declaring that Copernicus contradicted dogma.

The same is true about a statement by Ernst Brüche within the context of his commentary on the speech given by Cardinal Franz König at the conference of Nobel Prize winners in Lindau in 1968 (which in its day was sensational and was tellingly greeted with the laughter of the students who were present). In his remarks about this speech, Brüche recommends that the Catholic Church should "divorce" itself "from dogmas, of which the Assumption of the Virgin Mary into heaven is probably the most difficult obstacle to a rapprochement, because of its subject matter and because it was

[15] Walter Gerlach, "Zur Geistesgeschichte der Galileizeit," *Deutsches Museum* 32 (1964): 47. The essay gives evidence of other misunderstandings, too. The same drift is found also for example in Gerhard Hennemann, "Der Fall Galilei," *Zeitschrift für Religions- und Geistesgeschichte* 20 (1968): 61–69, who moreover does not even refer to historical dates correctly.

proclaimed only recently in our era."[16] As the context suggests, the learned physicist understands Mary's being taken up into eschatological perfection as a "miracle of nature"—a grotesque misunderstanding which even a modest effort to become properly informed would have prevented.

Joachim Otto Fleckenstein looks at the Galileo case from another perspective when he compares the proceedings of the Curia against Galileo with the policy of the Soviet authorities with regard to scientists who did not follow the party line, and writes, "In Rome they softened practically everybody up. In the Second Rome (Byzantium) we seldom find cases like this, although in the Third Rome (Moscow) we see them constantly."[17]

This brings into our field of view now the whole set of problems posed by the Inquisition; after all, Galileo was tried by that tribunal. The accusation of moral corruption, which overcame all stirrings of conscience on the part of both the accusers and the accused, extends to the Church and Christianity in general: "Under the pretext of being guardians of the truth, crimes were committed in the name of Christianity by the official leadership of the churches, which led to the extermination or dismissal of persons whose orthodoxy was doubted."[18]

Now the encyclical *Humanae vitae* is drawn into the debate too, in that many characterize this dogmatic papal letter by Paul VI from the year 1968 as a modern case parallel to the condemnation of

[16] Ernst Brüche, "Das Angebot des Kardinals: Aus der Eröffnungssitzung der 18. Lindauer Nobelpreisträgertagung," *Physikalische Blätter* 24 (1968): 362–363. Also Norbert A. Luyten, "Naturwissenschaft und katholische Kirche," *Freiburger Zeitschrift für Philosophie und Theologie* 17 (1970): 428–441.

[17] Joachim Otto Fleckenstein, *Naturwissenschaft und Politik von Galilei bis Einstein* (Munich: Verlag Georg D.W. Callwey, 1965), 51.

[18] Johannes Hemleben, *Galileo Galilei in Selbstzeugnissen und Bilddokumenten* (Hamburg: Rowohlt Verlag, 1979), 9.

Galileo. The common element, supposedly, is that the Church thereby disregarded well-known, guaranteed findings of the natural sciences. Whereas the one case was an attack on the individual's freedom of thought, the other case involved an intrusion into the human freedom to decide. "Instead of refraining from a judgment in this question, the Pope disregarded the reality of imminent overpopulation and famines, and showed instead that the Successor of Peter is ultimately still entangled in the mindset that condemned Galileo."[19]

These postulates then lead to the conclusion that is drawn from the Galileo case by the majority of those who are pledged to modern scientific thought: "It is a shame, but the Catholic Church and the modern natural sciences are worlds apart."[20]

An impressive example of this was the reaction to the article "Galileo was not a Martyr" penned by Gerhard Prause, which appeared in the weekly magazine *Die Zeit* dated November 7, 1980. The question that led off the article and neatly captured the gist of it, "But was it really a misjudgment that was handed down in the Vatican on June 22, 1633?" unleashed a veritable storm of indignation.[21]

Thereupon the magazine printed a whole series of letters to the editor, eighteen in all, not one of which was willing to evaluate the author's presentation objectively.[22] On the contrary, one reader asked, "Who is this Gerhard Prause, who shamelessly violates the moral code of scientific methodology, who trivializes the efforts made in developing scientific hypotheses and theories, and proves to be incapable of mourning with those poor, cruelly tortured victims

19 Brüche, "Das Angebot des Kardinals," 363.

20 Ibid.

21 The same gist was found in a presentation that first appeared in 1966: Gerhard Prause, *Niemand hat Kolumbus ausgelacht, Fälschungen und Legenden der Geschichte richtiggestellt* (Düsseldorf-Vienna: Econ Verlag, 1976), 173–192.

22 Letters to the Editor in *Die Zeit*, 48/21, November 1980.

of conscience?" Another speaks about Galileo as a "victim of terror," and Prause's representations are accused of being "lazy tricks that are easy to see through."

Another letter raised the specter of "natural science gagged by theology" and blamed the Church for "arrogating inappropriate competence." Another reader asked, "Didn't anyone at *Die Zeit* clench their fists at this?" And finally someone pilloried "the Catholic intention, which still exists even today, to suppress science and humanity with a feeble-minded (pardon the expression) theological construct" as a "challenge to human beings and citizens."

Recently, this long-running tradition off the beaten track of sober, serious research, this anticlerical, rather journalistic treatment of the Galileo case, defined by prejudices, has found expression once again. For example, Albrecht Fölsing writes about Galileo's recantation:

> We do not know what makes us tremble more: Galileo's servile, ingratiating willingness to deny himself completely and submit, or the terrible brainwashing capabilities that the General Commissioner Maculano practiced so masterfully that he, as a virtuoso of psychological conditioning, can serve as an almost unsurpassable example to the thought police of all lands and times. The destruction of a human being's self-respect within a few days without applying actual physical violence remains an achievement of fascinating repulsiveness.[23]

Such language condemns itself. It is far more surprising, of course, that even highly regarded historians of indisputable academic caliber

[23] Albrecht Fölsing, *Galileo Galilei — Prozess ohne Ende, Eine Biografie* (Munich-Zürich: R. Piper, 1983), 144.

still prove, especially in their presentation of the consequences of the Galileo trial, to be indebted to prejudices whose roots lie in an uncritical enthusiasm for progress and in nineteenth-century anticlericalism. The playbook still includes a colorful lament about the collapse of the natural sciences in Catholic Europe, especially in Italy, which the measures taken against Galileo allegedly caused.

Besides the fact that such a claim has no basis whatsoever in the historical record, which instead proves exactly the opposite, such judgments show also a remarkable lack of critical thought on the part of such historians about their own intellectual-historical and scientific-theoretical presuppositions. The modern positivism and the scientific optimism of the scientific-technological age are not examined critically by these authors, much less the tendency to absolutize the personality of the researcher, which corresponds to the individualism of the modern era.

In terms of these presuppositions, then, judgment is passed on Galileo's critics and on the Church.[24]

The very vehemence of such reactions, however, vividly shows the depth of this trauma which was left on society, not so much by the Galileo case itself in its time as by the presentation of it in nineteenth- and twentieth-century historiography.

Last but not least, in this context, are the legends that have sprung up luxuriantly around Galileo, which express the general judgment about his case. The most famous of them deals with the oft-quoted *E pur si muove*—"And yet it moves"—which Galileo supposedly said while stomping his foot defiantly as he left the building of the Inquisition after his recantation.

This, however, is a fable that was already being marketed in eighteenth-century pamphlets but must have originated earlier. In the year 1911, indeed, this sentence was discovered during the restoration of a

[24] See ibid., 27–28.

painting that dates back to 1643 or 1645 and should be attributed to Murillo or to his school, which shows Galileo in the prison of the Inquisition—the development of legends therefore started that early. In the picture the prisoner points to the sentence. In the sources for Galileo's biography, nevertheless, there is no trace of it.[25] The prison-legend served as the subject for later painters, too. And they could even rely on scientific literature.[26] No doubt it was rhetorically very effective when the main speaker at the Leibniz Jubilee in 1908 could say that "the new science developed from the imprisonment of Galileo Galilei, and its '*E pur si muove*' finally prevailed even over the jailers."[27] But it is not true. Galileo was sentenced *ad formalem carcerem,* in other words to a "*pro-forma* jail." There can be no talk about arrest and captivity, even during the trial. The days that he had to spend in the palace of the Inquisition (April 12–30, 1633), as well as the brief hour on May 10 and the time from June 21 to June 24 for the concluding hearing and recantation, he spent in the residence of the *fiscalis,* a rather high-ranking official of the Inquisition, who had set aside several rooms for him, where his servant could attend to him and bring meals to him from the kitchen of the Florentine envoy. During the remainder of his stay in Rome, Galileo was a guest of the Florentine ambassador in the Villa Medici on the Pincio near the Church of Santa Trinità dei Monti. Then his "imprisonment" continued in the episcopal palace of his friend Ascanio Piccolomini in Siena and also in his own villa in Arcetri near Florence.[28]

[25] See Georg Büchmann, *Geflügelte Worte* (Berlin: Hauder & Spenersche, 1972), 629–630; Adolf Müller, *Der Galilei-Prozess* (Freiburg im Breisgau: Werdersche Verlagshandlung, 1909), 160–161; Stillman Drake, *Galileo at Work: His Scientific Biography* (Chicago and London: Dover Phoenix Editions, 1978), 356–357.

[26] See Müller, *Der Galilei-Prozess,* 162–163.

[27] Ibid., 163.

[28] See ibid., 163–164.

Galileo was not incarcerated. Neither was he tortured. The acts of the trial do speak about an *examen rigorosum* — strict hearing — but this referred to the *territio verbalis,* the threat of torture in words, which however could not have made a great impression on Galileo, since it was generally known that persons over sixty years of age were never subjected to that cruel procedure. The threat was a procedural formality.[29]

Finally, another impressive scene is depicted, in which a half-naked Galileo, clothed only with a penitential shirt, reads aloud the formula of his recantation. But scholarship has long since pronounced its judgment on that, too.[30]

Still, the aforementioned legends demonstrate how an anticlerical spirit of the age embellished the historical picture of the Galileo case for the widest possible circles. Given this finding, it would be too little to speak only about a tenacious historical cliché. What becomes clear from the emotional component of the positions just listed is instead the existential dismay of a society marked by natural scientific thought, which thinks that it can grasp all of reality with its purely scientific categories.

Before we can return to a new, productive encounter between scientific and theological thought — not just for individual distinguished minds, which no era has lacked, but rather on the widest possible front — the Galileo trauma needs therefore to be healed.

Such a trauma exists, however, also in the case of the other partner to the dialogue — the theologians, the believers — although in a different way. They had to witness how the explosive course of progress in the natural sciences, especially in the

[29] See Hartmann Grisar, *Galileistudien: Historisch-theologische Untersuchungen über die Urtheile der römischen Congregationen im Galilei-Process* (Regensburg-New York-Cincinnati: F. Pustet, 1882), 89–94.

[30] See Müller, *Der Galilei-Prozess*, 166.

nineteenth century, first brilliantly rehabilitated Galileo, who had been censured by the Church in 1633, and then put the Church in the wrong. Moreover the Galileo question provided the proponents of anti-religious and anticlerical liberalism, which increasingly prevailed in the second half of the nineteenth century, with inexhaustible material for polemics: "On the anticlerical side, they hoped through the authentic story of Galileo to provide proof of the intolerance of faith toward science, or to expose in decisions the fallibility, either of the Church as such, or of the pope in his magisterial *(ex cathedra)* pronouncements."[31]

On the other hand, the task facing the Catholics was "to wrest from their opponents the territory confiscated by them with obvious injustice."[32] Therefore — and indeed for both camps — the matter involved considerably more than Galileo. This explains also the harshness of the apologetic tone in several stretches of the book by Hartmann Grisar, for instance, who in other respects produced probably the shrewdest and most thoroughgoing study on the Galileo trial, and in the works of other authors who also belong to the Jesuit order. Remarkably, the Galileo trial is not mentioned at all in the *Apology for Christianity (Apologie des Christentums)* by F.X. Hettinger, which went through many editions, while long discourses are devoted to the promotion of the natural sciences by the Church and to the accomplishments of Catholic researchers, many of whom belonged to the clergy.[33]

Such an obvious exaggeration of the problem went hand in hand with a certain Catholic studiousness with regard to the natural sciences, which the witty Lord Disraeli once described ironically as

31 Grisar, *Galileistudien,* 11.

32 Ibid.

33 See Franz Hettinger, *Apologie des Christenthums* V (Freiburg: Herdersche Verlagshandlung, 1900), 215–222.

follows (through a character in his novel *Lothair*): "[The Catholics'] clever men can never forget that unfortunate affair of Galileo, and think they can divert the indignation of the nineteenth century by mock zeal about red sandstone or the origin of species."[34]

Moreover those on the Catholic side strove to exonerate the judges by shedding light on the facts and the hidden difficulties, in particular also by reference to the historical circumstances of the trial. The judgment itself was no longer defended by any one of the authors. Besides Grisar, another Jesuit performed a great service with his historical research into the Galileo case: Adolf Müller, a German living in Rome, professor of astronomy and mathematics at the Pontifical Gregorian University and director of the observatory on the Gianicolo. He penned not only a two-volume textbook on astronomy (1904–1906), but also — among other things — monographs about Copernicus and Kepler, precious preliminary studies for his two slim volumes on Galileo,[35] in which he provided — based on the critical edition of Galileo's complete works that had meanwhile become available and also on the sources for the biography of him by Antonio Favaro[36] — a sober, objective, critical presentation of the events surrounding Galileo. He intended — and this was his apologetic method — to convince by objectivity, although he was not entirely successful in his tone and judgment. Ludwig von Pastor, who in his history of the papacy devoted an extensive presentation to the present topic, spoke sadly about the event and opined, "For theologians,

[34] Matthias Buschkühl, *Die Irische, Schottische und Römische Frage, Disraelis Schlüsselroman "Lothair" (1870)*, Kirchengeschichtliche Quellen und Studien 11 (Sankt Ottilien: EOS Verlag, 1980), 25–26.

[35] Müller, *Der Galilei-Prozess*, 166; Adolf Müller, *Galileo Galilei und das kopernikanische Weltsystem* (Freiburg im Breisgau: Herder, 1909).

[36] Favaro, *Le Opere di G. Galilei.*

the error of 1616 and 1633 was a constant warning for centuries beyond, which was also taken to heart. No second Galileo case occurred."[37]

Similarly the Swiss cultural historian Gustav Schnürer concluded in the 1930s, "Nevertheless the decisions made then remain deeply regrettable.... They created among Catholics a distrust that was too far-reaching, which hindered them from participating joyfully in these investigations," and which allegedly resulted in the "intimidation" of scientific research in Italy. "The Galileo case under Urban VIII still remains a painful rebuke, which can be lifted."[38]

On the occasion of the three hundredth anniversary of the death of Galileo in the year 1942, the physicist Friedrich Dessauer used even more dramatic language. He called the Galileo case a "Western tragedy" and gloomily entitled the chapter of his book depicting Galileo's final days "Failed"—while in other respects he strove for historical justice.[39] However, as a result of the rapid, progressive loss of the historical dimension in Western thought in the following years, all that remains of Dessauer's words are those about the tragedy and the failure of Galileo.

During that same year 1942, on the occasion of the three hundredth anniversary of Galileo's death, the Pontifical Academy of Sciences commissioned the Roman historian and professor at the Lateran University, Msgr. Pio Paschini, to write a biography. The manuscript that he submitted two years later recorded with painstaking exactitude every detail about Galileo's life that could be gleaned

[37] Ludwig von Pastor, *Geschichte der Päpste seit dem Ausgang des Mittelalters* 13/2 (Freiburg im Breisgau, 1929), 630. English edition: *The History of the Popes*, 40 vols. (available at the Internet Archive).

[38] Gustav Schnürer, *Katholische Kirche und Kultur in der Barockzeit* (Paderborn: Schöningh, 1937), 612, 763.

[39] See Friedrich Dessauer, *Der Fall Galilei und wir: Abendländische Tragödie* (Frankfurt am Main: Knecht, 1957).

from the sources, and it did not conceal the facts that are perceived today as painful. Leo XIII had shown objectivity, for example by opening the Vatican Archives, but in the influential Roman circles at that time that impartiality had been lost, to such an extent that the publication of Paschini's manuscript was not permitted. The book was not published until Galileo's four hundredth birthday in 1964, long after Paschini had died. The manuscript was now twenty years old, which meant that it did not take into account the literature that had appeared in those twenty years. Besides, Paschini himself, because of a lack of special expertise, had not satisfactorily reflected the state of the research. So it had become necessary to revise the manuscript so that it could hold its own in the academic world. This task was assigned to the Jesuit historian Fr. Edmond Lamalle, an expert in the field of the history of science.[40] A comparison of the manuscript with the printed edition of Paschini's *Galilei* shows that the editor altered the book more or less in more than one hundred passages. Conformity with the

[40] See the foreword to the book by Pio Paschini, *Vita e Opere di Galileo Galilei*, penned by the editor Edmond Lamalle, vii-xv, and also Roger Aubert, "L'état actuel de l'affaire Galilée," in *Colloque d'Histoire des Sciences*, Université de Louvain, Recueil de Travaux d'Histoire et de Philologie VI/9 (Louvain, 1976), 156–157. Concerning the obstacles to the publication of Paschini's book and the changes later made to it, see Pietro Bertolla, "La vicende del 'Galileo' di Paschini dall'Epistolario Paschini-Vale," in *Atti del Convegno di Studio su Pio Paschini nel Centenario della Nascita 1878-1978*, ed. Deputazione di Storia patria per il Friuli (Udine, 1979), 173–208. On evaluating Paschini as a biographer of Galileo, see Pietro Nonis, "L'ultima opera di Paschini: Galilei," in *Atti del Convegno*, 158–172—while on the other hand the abovementioned reservations apply. The critique by Paolo Simoncelli, "Inquisizione romana e Riforma in Italia," *Rivista Storica Italiana* 100 (1988): 95–97, of Lamalle's edition of Paschini is devoid of objective foundation, as was correctly underscored by Filippo Tamburini, "La riforma della Penitenzieria nella prima metà del sec. XVI e i cardinali Pucci in recenti saggi," *Rivista di Storia della Chiesa in Italia* 44 (1990): 128.

state of the research in the year 1964 had made this necessary, and there is no disputing that the work was improved thereby.

As a result of the change of mentality that had occurred meanwhile, Galileo's name was mentioned aloud even at the Second Vatican Council.[41] Interesting for our purposes is the context in which this happened: Cardinal Suenens was the one who, against the background of the celebration of Galileo's four hundredth birthday, took up a position with regard to the problem of regulating births at a plenary session of the council on April 30, 1964, during the deliberation of the much-debated Schema 13 about "The Church and the Modern World." He demanded a new draft of the text after consulting experts from all over the world, and exclaimed, "Let us follow the progress of science! I beseech you, my Brothers: let us avoid a new 'Galileo case.' One is enough for the Church."[42]

A similar line of argument was taken then on September 27, 1965, by the Indian Archbishop Eugene D'Souza, who expressed his displeasure with the Church's adaptation to the modern world, which in his view was too slow and too closed-minded. He quoted Suenens and continued:

> Yet since Galileo we have had — not to mention others — the Lammenais case, the Darwin case, the Marx case, the Freud case, and recently the Teilhard de Chardin case. To be sure, their works and the

41 See also Oswald Loretz, *Galilei und der Irrtum der Inquisition: Naturwissenschaft — Wahrheit der Bibel — Kirche* (Kevelaer: Butzon & Bercker, 1966), 143–151.

42 Johann Christoph Hampe, ed., *Die Autorität der Freiheit — Gegenwart des Konzils und Zukunft der Kirche im ökumenischen Disput* III (Munich: Kosel, 1967), 260.

> movements started by them were infected by certain errors. And yet these men were fighting for genuine values which our Schema is acknowledging today. Why then did they have to be condemned wholesale?[43]

These statements are hardly evidence of deeper historical knowledge.

In both instances, therefore, the name *Galileo* stands for the progress of science and human culture—in D'Souza's remarks, of course, in quite a different society.

From then on the shadow of Galileo hung over the Council Hall. The auxiliary bishop of Strasbourg, Léon Arthur Elchinger, conjured it up again on November 4 of that same year when he demanded a rehabilitation of Galileo conducted by the highest ecclesiastical authority. He saw this as a way of proving to the world that the Church was by no means anxiously and defensively opposed to modern culture, as many believed.[44]

In fact Paul VI complied with this initiative several months later in a rather indirect way, when he recommended during his visit to Pisa on June 10, 1965, that the Catholic people should imitate the faith of Galileo, Dante, and Michelangelo—he mentioned them in that sequence.[45] This carefully nuanced reference was then generally understood by the wider public.

And so it came to the very important formulation in the conciliar constitution *Gaudium et spes*, dated December 7, 1965, which says, among other things:

[43] Ibid., 34.

[44] See ibid.; Xavier Rynne, *The Third Session: The Debates and Decrees of Vatican Council II* (Farrar, Straus and Giroux, 1965).

[45] See Hampe, ed., *Die Autorität der Freiheit*, 330.

> By the very nature of creation, material being is endowed with its own stability, truth and excellence, its own order and laws. These man must respect as he recognizes the methods proper to every science and technique. Consequently, methodical research in all branches of knowledge, provided it is carried out in a truly scientific manner and does not override moral laws, can never conflict with the faith, because the things of the world and the things of faith derive from the same God.... Consequently, we cannot but deplore certain attitudes (not unknown among Christians) deriving from a shortsighted view of the rightful autonomy of science; they have occasioned conflict and controversy and have misled many into opposing faith and science.[46]

The footnote to this passage refers to the book by Pio Paschini about Galileo, which had just been published then. It is the only time that the council cites a contemporary scientific work.

The passage had been debated heatedly as it was being written. The suggestions that the Galileo case should be viewed in its historical context were not heeded, any more than the recommendation to say nothing at all about it. Two Council Fathers, who had given the latter advice, were of the opinion that such remarks would only manifest an inferiority complex of the Church with regard to

[46] Vatican Council II, Pastoral Constitution on the Church in the Modern World *Gaudium et spes*, in *Vatican Council II: The Conciliar and Post-conciliar Documents*, ed. Austin Flannery, O.P., new rev. ed. (Boston: St. Paul Books and Media, 1992), 935.

science.[47] What in fact became visible from these sentences by the council is the trauma that the Galileo case left behind in the Church, too. Ever since then, accordingly, there has been a new kind of eager apologetic, which of course no longer consists in defending those who were at one time in charge, but rather seeks to save face for the Church today by accusing the Church of the past and distancing oneself from it. But this too is apologetics and therefore a debatable way of confronting history.[48]

This tense relation to history was typically illustrated by a news item from the Deutscher Depeschendienst (DDP) news agency that was circulated in the press in late February 1981. It reported that Dominican Father William Wallace, who was teaching in Washington, had discovered in his research into Galileo's scientific work a large number of plagiarized passages.[49] This fact, concerning his unattributed borrowings from Kepler's publications, was rather unflattering for Galileo, but it had been known for a long time.[50] The point of interest now is that the scholar who discovered this plagiarism thought that in the same breath he had to neutralize the embarrassment of this fact by characterizing Galileo's behavior as common practice then, or even as flattering for the authors who had been plagiarized.

[47] See ibid. and commentary in the German edition *Lexikon für Theologie und Kirche: Das Zweite Vatikanische Konzil, Dokumente und Kommentare* III (Freiburg im Breisgau, Basel, Vienna, 1968), 3:387.

[48] A typical example of this is the above-cited survey of the actual state of the Galileo question by Roger Aubert and his evaluations.

[49] See *Süddeutsche Zeitung*, February 28, 1981, a review of W. A. Wallace, *Galileo's Early Notebooks: The Physical Questions, A Translation from the Latin, with Historical and Paleographical Commentary* (Notre Dame, IN: University of Notre Dame Press, 1977).

[50] See Müller, *Der Galilei-Prozess*, 53; Müller, *Galileo Galilei und das kopernikanische Weltsystem*, 115–118.

But anyone who has even a superficial idea of how jealously Galileo himself was intent on claiming that he was the first to make a discovery, and of how recklessly and bitterly he fought over it — for instance in the cases of the proportional compass theorem and of sunspots, to mention only the best-known examples[51] — knows how untenable this attempt to excuse him is. Nevertheless it shows — and this is why it is mentioned here — that plainly today, more than 350 years after the Galileo trial, it is still not always possible for a Catholic scholar to communicate simple, proven facts without fearing that he will touch open wounds.

We — and this means everyone interested in the relation between science and the Church — must again be able to speak impartially about the Galileo case, without aggression and without a guilt complex. This question addresses our relation to history, because the case is an event from the past and thus a component of our history. But what happens if a human being does not accept his own history, does not learn to live with it and on the basis of it? Psychoanalysts know the answer. Similar phenomena exist in the social dimension also. Hence both "communities," the republic of the learned and the Church, must learn to live with this history of theirs. Last but not least, it is the duty of secular and ecclesiastical historiography to enable them to do so.

Now first we should point out the necessity of understanding the Galileo case in terms of the presuppositions of his own time, but not of ours. Of course this demands considerable intellectual effort. It consists, first, of quite deliberately ignoring everything that has happened in the growth of our knowledge since the year 1633. An observer who is striving to comprehend the facts objectively must not only ascertain the facts exactly by means of the sources; he must also situate these individual facts in the intellectual, religious, cultural, and also political panorama of their time and consider the

[51] See Müller, *Der Galilei-Prozess*, 47–48; 89–90.

event that interests him not in isolation, but rather as it is interwoven in many ways in the overall historical situation. Moreover, as this is being done, we must recognize the historical contingency of our own intellectual and judgmental presuppositions. This means that we must take off our glasses, which are tinted by the experience of a secularized, pluralistic, and permissive society, if we want to do justice to the historical reality of the Galileo case.

If we apply this method in the present case, then it will be possible to see Galileo and his judges, not from our standpoint, but rather with the eyes of a contemporary who is himself still standing in the middle of the stream of events and cannot yet know everything that appears self-evident to someone who is born later. To the extent that a historian succeeds in putting himself into the world of those whose actions and sufferings he strives to research, he will also succeed in understanding why the course of events ran this way and not differently. Only this strict intellectual discipline, therefore, creates the prerequisites for historical understanding, which preserves us from anachronistic false judgments. Then a person may climb the lofty peaks that offer a panorama of world and human history and cause the observer to recognize major connections and to gain comprehensive insights.

The figure of the unhistorical, judgmental Pharisee, who sits on the lofty throne of his present day and blithely passes judgment on the past, therefore has its ridiculous features. He forgets that his own actions and failures and those of his contemporaries will one day become the object of judgment by later generations.

Part of the historical composition of mankind and society, and naturally also of the Church, is this progress in the knowledge of reality, as well as the fading or even the loss of insights that have already been gained. It was a long way from the production of the first stone tool to the landing on the moon, and the fact that the spark of genius does not usually jump from one to the many with lightning speed often made this and other paths so long, steep, and wearisome. Not infrequently the interval between the first glimmering of a thought and certain knowledge extends over several generations.

Anyone who keeps this in mind will be rather careful in passing judgment, even on Galileo and his judges, and because he is aware of the limitations of his own insight and of his own judgment, he will go to work with intellectual modesty. But then the treasury of human experience opens up to him, and history is nothing short of that; when it is understood, it does not hesitate to serve as "*vitae magistra,*" the "teacher of life," for natural scientists and theologians.

Chapter 2
The History

Beginnings in Pisa and Padua

When Galileo Galilei was born on February 15, 1564, in Pisa, not even four weeks had passed since Pope Pius IV had ratified the Council of Trent by the bull *Benedictus Deus*.[52] After the catastrophic incursions of Luther's and Calvin's Reformation, this council brought a new beginning of religious and ecclesiastical life in what remained of Catholic Europe, which, nourished by the deep sources of a purified and renewed faith, brought forth that "miracle of Trent," as the intellectual-religious-cultural flourishing that proceeded from it still appears to the historian today. In that same year Michelangelo and Calvin died and Shakespeare was born. The market in Frankfurt offered the first catalogue of books for sale.

Galileo's childhood years saw the rebellion of the Netherlands, the tragedies of Mary Stuart and of Don Carlos, and in 1571 the seven-year-old son of a cloth merchant joined in the jubilation over

[52] Since the biographical facts about Galileo and those of the history of his time are hardly controversial, they are generally not documented specifically. Instead, we refer once and for all to Paschini. Here the biographical details are presented with the desirable breadth. For Galileo's scientific biography, we refer the reader to the aforementioned book by Drake. See also Elio Gentili, *Bibliografia Galileiana, fra i due centenari (1942–1964)*, Hildephonsiana, Collana di studi teologici e religiosi 8 (Milan: La Scuola Cattolica, 1966).

the victory of the Christian fleets over the crescent moon at the Battle of Lepanto. The Huguenot wars raged in France, and in 1588 England conquered the Spanish Armada of Philip II.

In this world, in the politics and profession of faith, the mysticism and the might, the heroism and the art that were blended together in the pathos of the Baroque style, the young Galileo Galilei grew up. From his father, who was an important musician and theoretician of music, the son inherited a rich intellectual endowment. However, the Florentine patrician Vincenzo Galilei was not blessed with earthly possessions; he lived at that time as a cloth merchant in Pisa.

From him Galileo receive his first instruction, which was followed immediately by a thorough introduction to Latin and Greek literature; Galileo's contemporaries marveled at his intimate knowledge of them. The boy's stay with the Benedictines in Vallombrosa was significant for his further education. His intention to enter the community there and to become a monk was nevertheless thwarted by his father. According to a report by a contemporary, the Abbot of Vallombrosa, Vincenzo took his fourteen-year-old son out of the monastery on the pretext that an eye ailment required treatment. A career as a physician was to gain for the son what the father had so far pursued in vain: wealth and influence. For this reason he sent the seventeen-year-old Galileo to the University of Pisa, where the young man attended lectures on medicine and philosophy.

He did not complete those studies, however. Rather he joined the students of the renowned mathematician Ostilio Ricci, who instructed the pages in the court of the Grand Duke in Florence, to which the Galilei family had returned in 1574. Here he became acquainted with Euclid, Archimedes, and Pythagoras. That gave him access to the world in which he himself was to become one of the great figures.

His first works contained his first discoveries, and even back then Galileo gave signs of what would later be a strongly pronounced trait in his character: in December 1587 he had four witnesses confirm his authorship of his observations concerning gravity.[53]

Then for the first time he directed his footsteps toward Rome, where he became acquainted with the Jesuit Christopher Clavius, an important astronomer and mathematician and the father of the Gregorian calendar.[54] Three notebooks of Galileo written in Latin also date back to the years around 1590; the first two contain transcripts from lecture notes of young Jesuit professors of the Roman College, which Galileo then used as the basis for his teaching in Pisa.[55] This led to further scientific contacts, and after the interlude of a *lettura* (a term as a lecturer) at the University of Siena, his patrons created a professorial chair in Pisa for the twenty-five-year-old Galileo, despite his irregular course of education and his lack of an academic degree. The recommendations of the well-known mathematician Marchese Guidobaldo del Monte and of his brother, Cardinal Francesco, secured for Galileo an acquaintance with the Grand Duke Ferdinand and with Giovanni de' Medici. Here in Pisa Galileo followed in the footsteps of his great predecessors Tartaglia and Benedetti on the path that led to the formulation of the laws of gravity. What enabled Galileo to do this was the ingenious combination of experimental measuring and weighing with mathematics. This introduced the development of modern natural science — and the intellectual-societal

[53] See Favaro, *Le Opere di G. Galilei,* 1:183.

[54] Concerning Christoph Clavius, S.J. (1537/38–1612), see Edmond Lamalle, in *NDB* 3 (1957): 279.

[55] See William A. Wallace, *Galileo and His Sources: The Heritage of the Collegio Romano in Galileo's Science* (Princeton, N.J.: Princeton University Press, 1984).

process made possible and determined by it; we today must grapple with the provisional results of this process.

After the death of his father Vincenzo in 1591, Galileo had to take care of his mother and siblings. His brother Michelangelo had not learned a trade that could have provided him with an income; he had been trained only as a musician. The resulting economic pressure — his sister Virginia had to be given in marriage with a dowry, too — caused Galileo to look for a more lucrative position. After disappointments and difficulties of many kinds, the Signoria of Venice finally decided in September 1592 to confer on him for four-to-six years a professorial chair in mathematics in Padua, which earned him three times the income he had in Pisa. In addition Padua, much more than Pisa, offered the opportunity to make income on the side by private instruction, which was quite customary. As soon as he gave his inaugural lecture on December 7, 1592, the new mathematician had an extraordinary influx of students. His years in Padua were also the time of Galileo's relationship with a Venetian woman, Marina Gamba, with whom he lived together from 1599 on. She bore him three children, whom he took with him in 1610 when he left Padua and Marina Gamba.

Besides his lectures in geometry, Galileo applied himself even then to astronomy. In this field, in keeping with the view of the day, he followed the teaching of Ptolemy, without taking note of Copernicus and Tycho Brahe. That would hardly have been possible, either, in the milieu of Padua, which was marked by a strict Aristotelianism. But when did Galileo come to the point of abandoning the current view and changing to the Copernican system? A first sign of this is a letter from the year 1597, in which he asks an old friend why he argues so decisively against the Pythagoreans and against Copernicus. In the previous year Kepler in Tübingen had published his *Prodromus* and had sent a copy of it to Galileo in Padua. The latter wrote in his reply that he himself had changed over to Copernicanism many years before but did not dare to publish treatises on this matter, but Galileo's recurring inclination to ensure that he was the first to announce a discovery admonishes us to interpret the statement cautiously. Significant also is the remark that he would certainly feel encouraged to publish his works if there were more authors who thought as Kepler did. But as it was, he wanted to wait.[56]

As late as the year 1606 he presented to his students in the *Trattato della Sfera o Cosmografia* the Ptolemaic system without the slightest hint of doubt or criticism. It is significant that in support of the teaching that the earth is the center of the universe he adduced those proofs from antiquity which Copernicus had long since refuted. Of course Galileo ought to have grappled with this, and also with the teaching of Tycho Brahe, whom he likewise ignored; Clavius had long since celebrated the astronomer from Königsberg as a second Ptolemy.[57] This probably means that until then Galileo could not yet have dealt seriously with Copernicus. Presumably for the same reason he did not reply to Kepler when the latter asked him to inform him of his reasons in favor of the Copernican system — reasons which, Galileo insinuated, he himself had found. Galileo remained silent even after a second letter to this effect. It would take seven years for a new exchange of correspondence to begin. Furthermore, when we see that Galileo told Professor Giovanni Magini from Bologna that he did not know Kepler's book, it is no longer possible to suppress questions about Galileo's truthfulness.

[56] See Galileo to Kepler, Padua, August 4, 1597, in Favaro, *Le Opere di G. Galilei,* 10:67–68); Müller, *Galileo Galilei und das kopernikanische Weltsystem,* 13. On the genesis of Galileo's Copernican convictions, see Mario D'Addio, "Considerazioni sui processi a Galileo," in *Quaderni della Rivista di storia della Chiesa in Italia* 8 (Rome 1985): 11–17. On Johannes Kepler (1571–1630), see M. List, in *NDB* 11 (1977): 494–508.

[57] See Müller, *Galileo Galilei und das kopernikanische Weltsystem,* 6–9.

Such doubts are fed also by the way in which Galileo behaved during his first years in Padua in the dispute over priority concerning the proportional compass. Galileo did not invent this mathematical instrument, although he did develop it further. This proportional compass, which made possible both the reproduction of drawings to various scales and also the calculation of roots and percentages, was a precursor of our pantograph and also of today's slide-rule. Galileo composed a set of user's instructions for it, which he published in 1606.

A dispute ensued with a young man from Milan, Baldassare Capra, who the following year inexplicably published an extremely faulty translation of Galileo's work into Latin as his own, without mentioning Galileo's name even once.[58] It is quite understandable that Galileo was indignant about it and demanded reparations. The way he went about it, however, is surprising and disappointing: he set out to have the young man outlawed entirely, although the latter had offered him a complete retraction and public reparation.

After a big scene before the academic tribunal, which was painful even for the judges, the judgment was announced to the assembled student body amid trumpet blasts, and thereupon they proceeded to confiscate and destroy all copies of the miserable booklet. But that was not enough for Galileo; now he published a work of his own in which he also considered the details of an earlier astronomical work by Capra. The contents are very interesting, although one historian who is an expert astronomer says that Galileo was wrong. What is surprising here is the vocabulary that Galileo employs in this work. For example he addresses Capra's teacher, the Frankish Simon Marius from Gunzenhausen, as a jealous enemy, a diabolical advisor and misanthrope, who like a snake darts

[58] On Baldassare Capra (1580–1626), see G. Gliozzi, in *DBI* 19 (1976): 106–108.

his biting, lying tongue all around, ravenous to injure the honor of others, especially of him, Galileo.[59]

But what Galileo accused his opponent of doing, he now did himself. Although as late as 1606 he had indicated that the proportional compass was an instrument invented by someone else which he had improved, now in his written defense he declared that it was exclusively his invention.

These traits of the great scholar cannot be overlooked if we want to explain and understand his later fate.

The telescope and the consequences

The invention had been in the air.[60] Eyeglasses had been in use since the thirteenth century, and from 1580 on the Neapolitan Giambattista della Porta applied himself to scientific optics; he published his findings in 1589 in his *Magiae naturalis* and in 1593 in *De Refractione* (on refraction). Kepler did further theoretical work in this field. We probably have a Dutchman to thank for the fact that these theories were overtaken by praxis and a telescope was produced: Zacharias Janssen from Middelburg. There a glass industry had been established, introduced by Italians, and there is an admittedly unverifiable indication that a telescope built in 1590 was also brought from Italy to Middelburg, which Janssen used as a model to produce his instrument in 1604. However, one of Janssen's compatriots, Jakob Adrianszoon Meetsius from Alkmaar, also claimed this invention as his own. Be that as it may, in 1608 the first telescope appeared at

[59] See Adolf Müller, *Galileo Galilei und das kopernikanische Weltsystem*, 32-38.

[60] See Vasco Ronchi, "Storia del Cannocchiale," in *Miscellanea Galileiana* 3:725–848. The author was the head of the research institute established in Galileo's villa. See Albert Van Helden, "Galileo and the Telescope," in Paolo Galluzzi, ed., *Novità celesti e crisi del sapere: Atti del Convegno Internazionale di Studi Galileiani*, Supplemento agli atti dell'Istituto e Museo di Storia della Scienza Anno 1983, fasc. 2 (Florence, 1983), 149–158.

the Frankfurt Market, and from there this extremely important instrument for navigation and warfare started its victorious advance around the world.

Until then Galileo himself had never really dealt with problems of optics. However, from 1602 on he appears to have been occupied with optical lenses, and according to his own account, when he heard the news about the invention of the telescope, he set about building one for himself. That happened, then — as he says in *Sidereus Nuncius* — around ten months before its appearance, at any rate in the year 1609. It was the result not of theoretical deliberations and calculations, but of his practical skill. A contemporary account gives an idea of what a sensation this telescope caused among the leading circles of Venice, when it was set up there on the towering campanile of St. Mark's Basilica and brought to everyone who looked through it the surrounding cities and also the islands and the ships on the lagoon close enough to grasp. The Signoria was amazed, and with its practical sense it rewarded the builder with an annual income of 1,000 gulden for life.

The decisive step for the subsequent course of events, however, was that Galileo no longer directed his telescope only at the lagoon, but also to the starry skies at night.[61] This in fact began a new epoch of astronomy. "Renouncing earthly observations, I turned to heavenly ones. Then I first saw the moon as close as if it were only one earth's diameter away. Then I observed frequently and with incredible pleasure fixed stars and planets."[62]

[61] The court astronomer from Ansbach, Simon Marius (1573–1624), probably conducted astronomical observations by telescope some time before Galileo, in the course of which he discovered the moons of Jupiter in late November 1609. See Eckhard Pohl, "500 Jahre Astronomie in Franken," *Jahrbuch des Historischen Vereins für Mittelfranken* 95 (1990/1991): 89–90.

[62] German translation from Müller, *Galileo Galilei und das kopernikanische Weltsystem*, 47.

And now he describes his observations, first those of the moon. Its surface was quite unlike what the philosophers taught: it resembled that of earth, displayed valleys, hills, and mountains, and even dark surfaces reminiscent of seas. It was particularly impressive that among the Pleiades, for example, of which only six stars were known to date, more than forty more became visible through the telescope, and the Milky Way revealed itself now to the fortified eye as a collection of countless stars of greater and lesser magnitude.

Of even more far-reaching importance, however, was the discovery of the moons of Jupiter. Galileo reports:

> It was on January 7 of the current year 1610, during the first hour of the night, when in my telescopic observations of the starry sky I hit on the planet Jupiter. To my amazement I saw three brightly shining little stars, two to the east and one to the west of the planet, standing in an almost straight line parallel to the ecliptic. Naturally I took them for fixed stars and therefore paid them no further attention. On the following evening I lit upon the same object, but noticed now an altogether different constellation. The three little stars now all stood on the west side of Jupiter and closer to each other than the first time.[63]

Further observations determined that the planet Jupiter is accompanied by four moons, which circle it at unequal distances and different speeds. But this proved that a planet, itself circled by moons, certainly can revolve around the sun together with its moons. In connection with this Galileo also discovered Neptune, which of course he did not

[63] German translation from ibid., 49.

consider a planet. He observed it so exactly in the following years by means of an extraordinarily precise measuring tool that he had constructed, that with his seventeenth-century measurements he calls into question those of today's astronomy.[64]

Galileo presented to the educated public the astronomical discoveries that he had made with the telescope in a work published in 1610, which he called *Sidereus Nuncius* — Starry Messenger. The forty-eight-page treatise was — apart from the little work about the proportional compass — Galileo's first scientific publication. Still, at the time he was already forty-six years old. The title page of the volume alone,[65] which had appeared in the printing office of Tommaso Baglioni in Venice, arouses eager expectation, for it promises that "great and utterly marvelous spectacles" are offered in it to everyone, but in particular to philosophers and astronomers: discoveries that Galileo Galilei, a Florentine patrician and mathematician of the university in Padua, recently made by means of the telescope that he had invented [!]. They concerned the surface of the moon, the countless cloudy fixed stars of the Milky Way, and in particular the four moons of Jupiter, to which their first discoverer decided to give the name "*Medicea Sidera*" ("the Medici stars").

One modern author remarked that to our ears this announcement sounds a bit hyped,[66] yet Galileo was just employing the style of his day — which of course he handles in a masterly way. To this observation we can only add that the title page also contains the plain falsehood that Galileo invented the telescope.

64 See Stillman Drake and C. T. Kowal, "Galileis Beobachtungen des Neptun," *Spektrum der Wissenschaft* (February 1981): 77–89.

65 Reproduced in Accademia dei Lincei, *Galileo Galilei: Celebrazioni del IV. Centenario della Nascita* (Rome: Bardi Edizioni, 1965), illustration 16.

66 See Hemleben, *Galileo Galilei in Selbstzeugnissen*, 48.

This publication had a tremendous effect. Neither Copernicus nor Kepler with their works caused such an "outburst of emotions" on the part of the public as did Galileo's *Starry Messenger*.[67] In our day its effect should perhaps be compared with the experience that millions had when they followed on the television screen the first landing of a man on the moon.

More important, however, was the fact that in this publication Galileo repeatedly made it known that he considered the Copernican system true. In his dedication of *Starry Messenger* to Duke Cosimo II, he spoke clearly about the movement of Jupiter together with its moons around the sun, which he called the center of the universe. Twice he mentioned that he wanted to give a thorough, detailed description of cosmology soon in a work of his own. Of course, more than twenty years would pass before he was able to do this.

Recognition and Opposition in Florence and Rome

Galileo sought to get away from Padua. Certainly, he had achieved there his first great successes, and now, after the sensational discoveries in the firmament, the influx of auditors forced him to transfer his lectures to the largest auditorium, which held a thousand students. Yet there was no lack of opposition, either, for "modesty was not Galileo's virtue."[68] He himself complained bitterly about the envy and jealousy of his enemies. Yet he also entangled himself in contradictions when he wrote on the one hand that he had not only silenced his toughest opponents but also convinced them so thoroughly that they had publicly declared themselves his followers; yet a few months later wrote the exact opposite

67 See Arthur Koestler, *Die Nachtwandler: Das Bild des Universums im Wandel der Zeit* (Bern-Stuttgart-Vienna: Scherz, 1959), 373. English edition: *The Sleepwalkers: A History of Man's Changing Vision of the Universe* (Penguin Books, 1990).

68 Hemleben, *Galileo Galilei in Selbstzeugnissen*, 48.

to Kepler — namely that the most influential professors at the university stubbornly refused to look at the moon and the planets through the telescope, as he had often invited them to do.

Galileo was drawn to Florence. He had always maintained good relations with the court there, and the chancellor Belisario Vinta was his correspondent and friend, with whom he could speak with complete candor. Vinta was also the one who was working on Galileo's return to Florence.

Now the time for it seemed to have arrived. The great discoveries, the stabilized economic position in Padua, and the great prestige that Galileo already enjoyed created a favorable basis for starting negotiations with the court of the Grand Duke. By bestowing the name *Sidera Medice* on the four moons of Jupiter that he had discovered, he took a further step to secure for himself the favor of the Grand Duke's family. The same purpose was served by the extremely elevated dedication of the *Starry Messenger* to the exactly nineteen-year-old Cosimo II, who had succeeded his father as ruler in 1609. Moreover Galileo had good, indeed friendly, relations with him; after all he had been his teacher since 1609, when he had stayed in Florence in the summer.

So now Galileo seized the opportunity offered by the favorable hour and informed the majordomo of the court of Florence, Vincenzo Vespucci, in February 1609 of his desire to be summoned to Florence. His main motive: in a republic like Venice it was impossible to draw a municipal income without having to be at the disposal of the state and the citizens. But this was precisely what robbed him of the necessary freedom and the best hours for his research. Only a *principe assoluto*, an "absolute sovereign," could grant him a position that made this possible for him to the desirable extent. He could be of service to such a prince with all the inventions that he was making — and could produce even more of them, given the necessary leisure.

A journey to Florence, where the court waited with the utmost curiosity to be able to repeat now for itself Galileo's discoveries with the telescope, offered Galileo the possibility for further probing and negotiations. For although a heavy gold chain with a medal — it cost 400 scudi — sent in April by Cosimo II as a gift was extremely promising, many misgivings came over the young prince with regard to the opposition that the astronomer had in professional circles. Not until June 10, 1610, did he finally appoint him as "Principal Mathematician of the Studio in Pisa and Principal Mathematician and Philosopher of the Grand Duke of Tuscany."[69] The University of Pisa had to come up with his annual income of 1,000 scudi, although Galileo was not obliged to give lectures there. As early as June 15 he sent to the Venetian senate his resignation from his professorial chair in Padua, in which not even considerable offers of remuneration by the Signoria could keep him. On September 7 he left Padua and the mother of his children, taking them along with him, and arrived in Florence five days later. Galileo did not leave a good memory behind in Venice; the people were exasperated by his departure, and he himself never again returned to that city.

But what was the effect of the response to the *Starry Messenger* on the scientific world?

Through the Florentine ambassador Giuliano de' Medici, the imperial mathematician Johannes Kepler had received the *Sidereus Nuncius* immediately after its publication. As early as May 2, 1610, enthusiastic about his reading of it, he had composed a work in reply, in which he lavished extravagant praise on the Paduan astronomer. Since Kepler, in setting forth Galileo's results, took his own just-completed work *Astronomia nova* as the point of departure, he

[69] *Primario Mathematico dello Studio di Pisa e primario Matematico e Filosofo del Granduca di Toscana.*

advanced the scientific discussion. However, the fact that in doing so Kepler could cite his own ideas, which had already been expressed in writing, may have annoyed Galileo so much that for his part he did not even take note of the *Astronomia nova* — even though Kepler had made no secret of his admiration for Galileo. In particular he had emphasized that through the discovery of the moons of Jupiter, the Copernican system had gained a strong supporting argument.[70]

There was a similar response by the famous creator of the Gregorian calendar, the Jesuit Christopher Clavius, originally from Bamberg and a teacher of mathematics and astronomy at the Roman College. He was a considerably older man, and Galileo had been in contact with him since his youth. Now he too had received the *Starry Messenger* and hastened to procure a telescope for himself in order to verify Galileo's observations. It seems, however, that a scientific exchange did not develop until late 1610 or early 1611, after Clavius had heard about Galileo's observations of Saturn. According to them, Saturn consisted of three stars. This, Clavius wrote now on December 17, 1610, could not be observed there in Rome — to him the planet appeared to be oval in shape — and he added a drawing to this effect. Clavius concluded his letter with amicable, indeed cordial, words and encouraged Galileo to continue with his observations and to include other planets, too. A productive exchange of observations and ideas was thereby initiated.

Thus Galileo had met with the approval of two leading astronomers. But there were still the opponents, too. For example, he received from Paris the demand for a copy of the *Starry Messenger* by a doctor who also informed Galileo that the mathematicians there were decidedly opposed to him. At the imperial court in Prague, too,

[70] See Adolf Müller, *Johann Kepler, der Gesetzgeber der neueren Astronomie: Ein Lebensbild* (Freiburg im Breisgau, 1903; repr. Outlook Verlag, 2022), 97; M. List, in *NDB* 11 (1977): 500–501.

they were initially skeptical, until Kepler composed his written response to the *Sidereus Nuncius*. Nonetheless, Kepler too noted that Galileo had not yet come to terms with Copernicus or Tycho Brahe, nor with Giordano Bruno, nor with himself.

Galileo found a real opponent, and a dangerous one on account of his influence, in the person of the renowned Professor Magini of Bologna, who for all his formal courtesy took the lead among Galileo's adversaries. His motive was perhaps scholarly envy. In any case, he felt that the foundations of his hitherto firmly held convictions were being attacked. He declared Galileo's book a fraud, yet obtained for himself from Venice a telescope in order to get to the bottom of the matter. A voluminous correspondence with the leading mathematicians of Germany, France, Flanders, Poland, and England served his anti-Galileo propaganda. Along the same lines, a student in Bologna by the name of Martin Horky responded, but without Magini's erudition; he was a Czech, the son of a Lutheran pastor who had once been friends with Kepler. He too declared that Galileo's discoveries were baseless inventions, and that he, Horky, could observe nothing of the kind with his own telescope — that was the gist of the lampoon that appeared in the summer of 1610.

Even a Florentine, Francesco Sizzi, sharpened his pen against his compatriot. While he admitted that he himself had seen the so-called moons, he traced this "seeing" back to optical illusions. After that he turned to the "proof" that there are only seven planets, and that therefore other moving celestial bodies could not exist. Astronomers like Clavius and other Jesuit professors just laughed at this botched critique.

Now of course it is easy, against the backdrop of our current knowledge, to join in this laughter. However justice must be done to Galileo's opponents, too. On the one hand it should be pointed out that his theses were startlingly new and contradicted generally accepted convictions, so that skepticism should not be surprising. Moreover, in the background stood the problem of the mutual ordering of experience and theory, on which Galileo and the Aristotelians were of different minds.[71]

On the other hand it must be grasped that the telescopes available in those days were anything but exact instruments. Even Kepler found it impossible for months to see what Galileo had described. This may have been due, above all — and this goes also for the fiascos of other astronomers with the telescope — to the fact that the distance between the eyepiece and the object lens could not be changed, and thus the telescope could not be adjusted to the individual viewer's acuity of vision. Galileo himself seems to have considered this only in 1634.

Therefore it was not necessarily envy or "spite … that blinded their eyes"[72] when Galileo tried in vain to demonstrate the satellites of Jupiter to twenty scholars in Magini's house.

The journey to Rome

In March 1611 Galileo traveled to Rome. Twenty-four years had passed since he had first visited the Eternal City as a twenty-three-year-old. What awaited him there now was probably overwhelming. During the pontificate of the Borghese Pope Paul V a true explosion of architecture and art had begun. Carlo Maderno and Carlo Fontana were putting their own stamp on Baroque Rome. St. Peter's was nearing completion, the Quirinal Palace was being built, the Cappella Paolina was added to Santa Maria Maggiore, and work was proceeding on the

[71] See Paul K. Feyerabend, *Der wissenschaftstheoretische Realismus und die Autorität der Wissenschaften: Ausgewählte Schriften*, I, Wissenschaftstheorie, Wissenschaft und Philosophie 13 (Braunschweig-Wiesbaden: Vieweg & Teubner Verlag, 1978), 252–280.

[72] Müller, *Galileo Galilei und das kopernikanische Weltsystem*, 56–57.

Acqua Paola Fountain. The pope and the cardinals spent large sums on the restoration of old church buildings and the construction of new ones; the pope's nephew Scipione Borghese was outstanding among them as an art collector and patron.

In a religious respect, too, Rome experienced a flourishing in those years. The names Camillo de Lellis and Giuseppe Calasanzio, for example, illustrate the care for the sick and the youth to which clerical Rome was dedicating itself.

What attracted Galileo this time, however, was scholarly Rome, which was focused on the Collegio Romano of the Jesuits and the Accademia dei Lincei (Lincean Academy). In particular the Jesuits of the Roman College were occupied in those years with astronomy—it is enough to mention the names of Clavius, Christoph Grienberger, Odo van Maelcote, and Giovanni Paolo Lembo. Galileo must have been anxious to make contact with them. In the circles of the College of Cardinals, too, some were interested in astronomy—and not only as dilettantes. Galileo himself had sent telescopes to Cardinals del Monte, Montalto, Acquaviva, and Borghese; Odoardo Farnese wanted to have two.[73] They were all eager to see the famous explorer of the firmament and to hear about his investigations, when he arrived in Rome on March 29, 1611, well equipped with letters of recommendation from Cosimo II.

A never-ending series of invitations, receptions, and conversations commenced, in the course of which Galileo met with the leaders of Roman society. And over and over again they gathered around the telescope, sought, found, and marveled at Jupiter's moons and sunspots and the Milky Way, for example in the Gardens of the Quirinal Palace, to which Cardinal Bandini had invited an illustrious band of secular and clerical guests. The high point,

[73] See Pio Paschini, *Vita e opere di Galileo Galilei*, in *Miscellanea Galileiana* I-II, 142.

however, was an audience with Paul V who, although at the time had little interest in the starry firmament on account of the theological concerns caused by the development of the debate on grace, nevertheless honored the scholar from Florence with signs of extraordinarily high esteem. Contrary to all protocol he did not allow Galileo to speak with him on bended knee.

With a view to the future events, however, it is important to note that Galileo met with great approval precisely in the College of Cardinals, although not all his admirers shared his Copernican views.

But how did the academic world receive the astronomer from Florence? Apart from the fact that Galileo's first steps on Roman soil had led him to Clavius at the Roman College — we will speak later about their further meetings — he was received on April 25, 1611, by the Accademia dei Lincei at a solemn session and admitted as their sixth member.[74]

This "Academy of the Lynx," whose successor is to this day one of the most highly respected scientific societies, had been founded by Prince Federigo Cesi, who was personally less interested in astronomy than in botanical investigations.[75] Moreover they paid allegiance to the ideal of the polymath and dealt with all branches of science and knowledge. For the rest of his life Galileo proudly called himself a "*Linceo*."

Even more important, however, was the decision of the Roman College to hold in honor of Galileo a scientific *symposium* on May 13, during which Fr. van Maelcote praised Galileo in the presence of

74 See "Documenti Lincei e Cimeli Galileiani esposti nella mostra organizzata nella Biblioteca Accademica a cura di A. Alessandrini," in Accademia dei Lincei, *Galileo Galilei*, 145–225; Raffaello Morghen, "Galileo e l'Accademia dei Lincei," in ibid., 131–144. Also Richard S. Westfall, "Galileo and the Accademia dei Lincei," in Galluzzi, *Novità celesti*, 189–200.

75 About Federigo Cesi (1585–1630), see A. Ferrari, in *DBI* 24 (1980): 256–258.

many distinguished guests, celebrating him as a "very famous and most fortunate explorer of the stars." He confirmed his discoveries, and then recalled the latest findings — the sickle shape of Venus and Saturn. He read Galileo's letter to Clavius about the oval form of Saturn verbatim and with it also Galileo's statement that the discovery of the phases of Venus proves that the heliocentric system is the only correct one. Incidentally, the speaker remarked that he was content to present the facts, while drawing conclusions from them was a matter for his listeners.

Galileo's triumph was complete, so complete that Cardinal del Monte wrote on May 31, 1611, to the Grand Duke of Tuscany: "If we still lived in the times of the old Roman Republic, I surely think that we would have set up a statue of him on the Capitol in recognition of his accomplishment."[76]

Of course the decisive thing was that the Jesuit astronomers of the Roman university acknowledged with Galileo that — as the phases of Venus demonstrated — at least one planet moved around the sun, and thus the Ptolemaic system was no longer tenable. The conclusion that they drew from this was of course much more cautious than Galileo's: they turned first to Tycho Brahe and took a conciliatory position with respect to Copernicus. Galileo himself, in contrast, made no secret of his Copernican conviction. Although he explicitly took this position in none of his writings from those years, nevertheless he missed no opportunity to promote Copernicus in conversation. In doing so he may scarcely have realized that he came into contradiction with the generally prevailing conviction that the earth stood still, and provoked everyone who also defended it scientifically. He had no lack of good friends to point this out to him and

[76] German translation from Müller, *Galileo Galilei und das kopernikanische Weltsystem*, 72.

to admonish caution.[77] In particular, the mathematician and astronomer Msgr. Giovanni Battista Agucchi was the one who advised him emphatically in July 1613 that Ptolemy could be dismissed only on the basis of solid mathematical proofs.[78]

They would be all the more necessary, since now a man appeared whose hierarchical, scientific, and humane-religious rank necessarily lent the greatest possible weight to his position in advance: Cardinal Robert Bellarmine. He too was among those who had looked through the telescope together with Galileo as guests of Prince Cesi — whose fatherly friend Bellarmine had been since his youth — and of Cardinal Farnese. Now, on April 19, 1611, he presented five questions to his confreres Clavius, Grienberger, van Maelcote, and Lembo. He wanted to know whether these individually listed observations of Galileo corresponded to the facts. Only this, but precisely this. The answer followed within five days; it contained with minor differences the confirmation of Galileo's observations. Now it was a question of what conclusion should be drawn from these discoveries. The work of coming to terms with Copernicus began, and Bellarmine assumed the leading theological role in it.

The remainder of Galileo's stay in Rome, which ended on June 4, 1611, was very encouraging for him, but not necessarily for Copernicus, for while Galileo's observations may well have raised doubts about the Ptolemaic system, they did not prove Copernicus correct.

A modern historian comments on this:

> It seems to be true that Galileo, fully convinced of the truth of the Copernican system, naively thought that he could convince the Cardinal [Bellarmine], too, and

[77] See the statements compiled in Paschini, *Vita e opere*, 277–299.

[78] See D'Addio, "Considerazioni," 16–17. About Giovanni Battista Agucchi (1570–1632), see R. Zapperi and I. Toesca, in *DBI* 1 (1960): 504–506.

> with him the whole *Collegium Romanum,* after they had concurred with his astronomical observations. Perhaps in Bellarmine's careful and evasive answers he had not noticed at all the iron resistance that extended to all the logical conclusions that Galileo drew from his observations. By no means did Galileo succeed in changing the mind of the Cardinal, the crucial man in the Curia for these cosmological questions, with regard to the Copernican system. But the Cardinal surely noticed the intensity with which Galileo adhered to the teaching of the Cathedral rector in Frauenburg, which in the Inquisition's view was dangerous and in no case admissible.[79]

Shortly before Galileo's departure from Rome, on May 17, the Inquisition supposedly took an interest in Galileo's earlier life: "Thus his name was recorded in the acts of the Inquisition."[80] Apart from the fact that the alleged Roman inquiry in Padua, as to whether Galileo's name had been mentioned in a then-ongoing trial against the philosopher Cesare Cremonini, took place not on May 17, but rather earlier on February 17 (in other words even before Galileo came to Rome in the first place),[81] the lines just cited are based on the author's imagination, not on sources.[82]

79 Hemleben, *Galileo Galilei in Selbstzeugnissen,* 70. In saying this, Hemleben overlooks the fact that at that time the Inquisition had not yet dealt with Copernicus at all.

80 Ibid.

81 See Paschini, *Vita e opere,* 88.

82 In reality Bellarmine had inquired whether Galileo was involved in the proceedings of the Inquisition against Cremonini. The latter, however, was an opponent of Galileo, which Hemleben et al. do not seem to know! See Drake, *Galileo at Work,* 487n20; Favaro, *Le Opere di G. Galilei,* 19:275.

Dostoyevsky's Grand Inquisitor may have served as the model for the picture of Bellarmine that is drawn here, when the account goes on to say that through this meeting with him, who was substantially involved in the trial against Giordano Bruno, a "shadow" had fallen on Galileo's life "for the first time, which from then on was to grow ever darker until his death."[83]

Who was this cardinal?[84] Roberto Bellarmino originated in the Montepulciano region of Tuscany; his mother was a sister of Pope Marcellus II. At the age of eighteen, in 1560, he became a Jesuit, and from 1570 on he worked as a theology professor and preacher in Louvain and Rome. He led the Jesuit Province of Naples, became archbishop of Capua in 1602, and, after being created a cardinal in 1599, spent the rest of his life in Rome, where he died in 1621. Of his numerous theological works, we should mention here only his famous three-volume *Controversies,* in which he grappled so thoroughly and so incisively with the teaching of the Reformers that in the Protestant sector of Europe it was forbidden to read this work and a number of "anti-Bellarmine professorial chairs" were established at universities.

We must not fail to mention that Bellarmine himself had come into conflict with the Inquisition. In volume 1 of his *Controversies* he had rejected the pope's claim to possess temporal sovereignty outside the State Church, which is why Sixtus V had the work placed on the Index of Forbidden Books. Only the fact that Sixtus V died shortly before the Index was printed made it possible for his successor, Urban VII, to revoke this erroneous decision. Bellarmine

[83] Hemleben, *Galileo Galilei in Selbstzeugnissen,* 69.

[84] See James Brodrick, S.J., *Robert Bellarmine, Saint and Scholar* (Westminster, MD: Newman Press, 1961). See also the short but sympathetic appreciation of Bellarmine by Cesare Vasoli, "'Tradizione' e 'Nuova Scienza': Note alle lettere a Cristina di Lorena e al P. Castelli," in Galluzzi, *Novità celesti,* 92.

therefore knew from experience about the occupational hazard of a theologian and could therefore have sympathy with Galileo's situation. Moreover, Bellarmine was an unworldly, deeply pious, and humble religious, who even as a cardinal lived a modest and secluded life. His great personal goodness was praised even by his contemporaries, and in 1930 he was declared a saint.

Now as far as astronomy is concerned, Robert Bellarmine could certainly serve as a competent dialogue partner. He had not only studied and taught astronomy, but also conducted later a scientific exchange with Clavius and had promoted astronomical studies as Rector of the *Collegio Romano*. It is particularly worth mentioning that he had based his own lectures about astronomy on the work by Alessandro Piccolomini, *Della sfera del mondo* (Venice, 1540), which among other things presented the hypothesis of the earth's movement.[85]

This, then, in broad strokes, was Cardinal Bellarmine, who moreover laid down no rules at all with regard to cosmology, and indeed was much more flexible than all of Galileo's philosophical and scientific opponents, as will be shown later.

Copernicus and the Church

In order to grasp in their historical significance the theological discussions that are now beginning, a flashback to the year 1543 is necessary, when the pioneering work of the cathedral rector in Frauenburg, Nicolaus Copernicus, *De revolutionibus orbium coelestium* (On the revolutions of the heavenly bodies) had first appeared in print. After an almost ten-year delay, the author had allowed the manuscript to

85 See Ugo Baldini, "L'astronomia del Cardinale Bellarmino," in Galluzzi, *Novità celesti*, 293–305; George V. Coyne and Ugo Baldini, "The Young Bellarmine's Thoughts on World Systems," in *The Galileo Affair: A Meeting of Faith and Science*, Proceedings of the Cracow Conference 24 to 27 May 1984, ed. G. V. Coyne, M. Heller, and J. Zycinski (Vatican City, 1985), 103–110.

be snatched from him; when it appeared in print, he had already been dead for several months.

Copernicus's teaching about the stationary sun and the earth moving about it was however not as revolutionary as is often assumed.[86] The Parisian professor and later bishop Nicolas d'Oresme had already taught in the mid-fourteenth century that the earth rotates on its axis, and he was followed a century later in this by the great Nikolaus of Kues. Moreoever, a contemporary of Copernicus, the prelate Celio Calcagnini of Ferrara, published as early as 1520 a work with the title *Che il cielo sta fermo e la terra si muove* — "The sky stands firm and the earth moves."[87]

Copernicus was born in 1473 in the Ermland region in Poland; he studied at the Universities of Bologna and Ferrara, and he met with much approval in Rome in the year 1500 with his lectures to well-educated circles. Rome in the Renaissance was open to everything called science. For example, Clement VII had his secretary, Johann Albrecht Widmanstetter from Ulm, explain to him the teachings of Copernicus in 1533. Bishop Tiedemann Giese of Kulm and the Dominican Cardinal Nikolaus von Schönberg (d. 1537) were the ones who urged Copernicus to publish his findings. When he did, Pope Paul III accepted the dedication of the work, and it had unhindered access and unbiased acceptance in the Catholic world. At the University of Salamanca, for instance,

[86] See the numerous investigations by Anneliese Maier in connection with her book *Die Vorläufer Galileis im 14. Jahrhundert* (Rome: Edizioni di Storia e letteratura, 1949). The very restricted sense we can speak about "precursors" is explained by Ernest A. Moody, "Galileo and his Precursors," in *Galileo Reappraised*, ed. Carlo L. Golino (Berkeley: University of California Press, 1966), 23–43. On Copernicus, see H. Schmauch, in *NDB* 3 (1957): 348–355.

[87] About Celio Calcagnini (1479–1541), see V. Marchetti, A. Ferrari, and C. Mutini, in *DBI* 16 (1973): 493–498. About Nikolaus of Kues, see 281.

astronomy lectures were given in 1561 according to either Ptolemy or Copernicus, but from 1594 on the latter alone ruled the professorial chair in astronomy there. The later Cardinal Pazmány, who was famous for his work on behalf of the Catholic reform of Hungary, presented the Copernican system as a professor in Graz and advocated unchallenged the view that nothing against Copernicus can be concluded from the well-known passages of the Bible. Moreover his calculations were the basis for the calendar reform of Gregory XIII. At around the same time — in 1581 — Bishop Martin Kromer of Ermland erected in the cathedral in Frauenburg a marble plaque commemorating Copernicus as the "great astronomer and renovator of the science of astronomy."[88]

Yet Copernicus himself had already counted on opposition; after all, he wrote in the foreword to his book, "If mere babblers ignorant of all mathematical knowledge should nevertheless presume to judge and dare to find fault with and attack this book of mine by deliberately twisting some passage or another of Sacred Scripture, then I will pay them no attention, but will despise their judgment as rash."[89]

The expected opposition came — not from Rome, however, but from Wittenberg.[90]

Even before Copernicus's work had appeared in print, Martin Luther made this statement about him in his table talk on June 4, 1539:

[88] See Grisar, *Galileistudien*, 279–281; Freiherr von Pastor, *Geschichte der Päpste*, vol. 12 (1927), 212. See also the summary in D'Addio, "Considerazioni," 6–11.

[89] "Si fortasse erunt mataiologoi, qui cum omnium mathematum ignari sint, tamen de illis iudicium sibi sumunt, propter aliquem locum scripturae, male ad suum propositum detortum, ausi fuerint meum hoc institutum reprehendere ac insectari: illos nihil moror, adeo ut etiam illorum iudicium tamquam temerarium contemnam." N. Copernici Torinensis, *De revolutionibus orbium coelestium*, libri VI (Norimbergae, 1543), praefatio fol. 4 v.

[90] See Hans Blumenberg, *Die kopernikanische Wende* (Frankfurt am Main: Suhrkamp, 1965), 100–121.

> A new astrologer was mentioned, who tried to prove that the earth is moved and goes about, not the heaven or the firmament, sun, and moon; just as if someone sitting on a wagon or on a ship and being moved along thought that he stayed still and rested, while the earth and the trees went about and moved. But so it goes now: he who wants to be wise will not stand for what others do; he must do something himself; that must be the very best way of doing it. The fool is trying to turn the whole art of *astronomiae* upside down. But as Sacred Scripture shows, Joshua told the sun to stand still, and not the earth.[91]

Probably anticipating such resistance from Wittenberg, Andreas Osiander, the Reformer from the imperial city of Nürnberg, who in 1543 arranged for the publication of *De revolutionibus orbium coelestium,* added a foreword (without however signing it as its author), in which he presented the Copernican system as a mathematical working hypothesis, which facilitated astronomical calculations that did justice to the phenomena observable in the heavens. To this he added the explicit remark that such hypotheses by no means needed to be true, or even probable, as long as they served their purpose. Moreover, he opined, it is plain to see that astronomy does not know the causes of the heavenly movements at all and has no intention whatsoever of showing that things in heaven behave this way and not differently — instead it is a matter of calculations that are correct in themselves.

With this foreword Osiander made history, as we will see.

[91] *Martin Luthers Werke: Kritische Gesamtausgabe. Tischreden* I (Weimar, 1912), 419, or *Tischreden* IV (Weimar, 1916), 412–413.

Six years after the appearance of the book, Melanchthon opined in his *Initia doctrinae physicae* that the hypotheses of Ptolemy, which were confirmed by the testimony of so many centuries, must not be dismissed rashly. He opposes Copernicus directly and says that there are people who despite all appearances make up all sorts of foolishness about the movement of the heavenly bodies, whereas physical and biblical reasons stand in the way of such novelties in astronomy. Given the great darkness of the human mind, one should cling to the word of God, which is to be understood in its literal sense.[92]

The standpoint proclaimed in Wittenberg was adopted by the inhabitants of Tübingen also, whose opposition to Copernicus was felt by Johannes Kepler, too. On account of his adherence to Copernicus, he had to leave Tübingen, and he was never able to return. In 1597 his teacher Hafenreffer had advised him, "God forbid that you should try publicly to bring your hypothesis into harmony with Sacred Scripture; work, I beg you, as a pure mathematician and do not disturb the Church's peace."[93] From then on Kepler found his sphere of influence in the Catholic sector.

Tycho Brahe, likewise an "imperial mathematician" and a Protestant, was moved by the literal meaning of the Bible, which he saw as being in conflict with Copernicus, to develop his own cosmology, which while preserving the geocentric principle was supposed to improve on Ptolemy. From then on the development among the Catholic scientists continued in the footsteps of Copernicus. In the field of Protestant theology, however, this no to Copernicus persisted for a long time. Throughout the seventeenth and eighteenth centuries literary voices spoke up again and again along these lines. Indeed, even in the scientific-technological nineteenth century,

[92] See Blumenberg, *Die kopernikanische Wende,* 371–395 (especially about Melanchthon's importance in this connection).

[93] Grisar, *Galileistudien,* 125.

Protestant writers still fought against the heliocentric system in the name of the Bible. We mention here only the preacher at the Church of Bethlehem in Berlin, Pastor Gustav Knak, who as late as 1868 continued his attacks against Copernicus.[94]

The skepticism that Galileo's Copernican propaganda met with in the Catholic milieu had roots that were not primarily theological but rather philosophical — apart from the fact that the Ptolemaic system corresponded to everyday experience and in many of its details was connected with people's living habits. Medicine, too, considered particular positions of the moon and planets when bloodletting and in many other matters. Above all, however, astronomy and astrology were still closely enmeshed. Even politics and warfare were dependent on horoscopes, which of course were drawn up on the basis of Ptolemy. Even Kepler had difficulty resisting such demands.

Thus the vicar general of Padua, Paolo Gualdo, adequately expressed the consciousness of the world at that time when on May 6, 1611, he wrote to Galileo:

> So far I have met neither a philosopher nor an astrologer who would subscribe to your opinion that the world revolves, and theologians are much less likely to do so. Therefore consider carefully before you maintain and publish this opinion of yours as being true. A lot can be debated that cannot very well be taken as certainly true; especially when one goes against public opinion, which has existed so to

[94] See ibid., 286–289; Hermann Theodor Wangemann, *Gustav Knak, ein Prediger der Gerechtigkeit, die vor Gott gilt* (Berlin, 1879); Hermann Theodor Wangemann, *Zeugnisse aus dem Leben des theuren Gottesmannes Gustav Knak* (Berlin, 1879); Otto von Ranke, in *ADB* 16 (1882): 261–263.

> speak "*ab orbe condito*" [since the creation of the world].[95]

Since Galileo usually despised such advice, it was inevitable that opposition would develop — Galileo received the first news about it from Rome in December 1611 through his friend, the important painter Cigoli.[96]

The conflict makes its way

Meanwhile Galileo himself had already received an answer. No sooner had he returned home from Rome than he found himself in heated arguments with the philosophers of the Peripatetic school; his opponents were the scholars of Pisa. The point at issue, however, was not astronomical in nature but rather the relation between solid and liquid bodies. Whereas they maintained that ice is heavier than water, because ice is water solidified by cold, and what is solidified is heavier than liquid, Galileo disagreed. His observations had taught him that ice floats on water. This led in the autumn to his *Treatise on Bodies that Move or Float in Water* (1612). The real topic of debate, however, was not the problem of specific gravity, but rather scientific method. Galileo's opponents, who were usually called Aristotelians or Peripatetics, were not natural scientists in today's sense. Starting from the works of Aristotle, whom they honored as their highest authority, their epistemological method consisted chiefly in the interpretation of the teachings of the great Greek, which was to be undertaken according to the laws of logic. Galileo, in contrast, followed his observations and measurements, which he then subjected to mathematical calculation. With that — and this is the reason for his

95 Paschini, *Vita e opere*, 280; Favaro, *Le Opere di G. Galilei*, 11:100.

96 About Lodovico Cardi, nicknamed "il Cigoli" (1559–1613), see M. Chapell, in *DBI* 19 (1976): 771–776.

prestige in the history of science — he had put all natural science on a new foundation, which for the first time was capable of supporting the load, and thus had inaugurated a new era in the development of natural science.

The debate about these questions of principle was bitter. Arguments appeared in writing, and counterarguments followed on their heels, and the fact that Galileo wrote fluently, was a brilliant stylist, and also gifted in the art of satire, mockery, and irony, ensured his literary success as a publicist. Thus he propagated a new understanding of science, which is not content with the mere proof from authority but rather investigates things themselves with the senses, with measurements and weighing and calculations.

One opponent who crossed swords with Galileo in this debate would later bring the argument about cosmology to a crisis also: the Florentine Lodovico delle Colombe. Others took up positions at his side. Their attitude perpetuated the impulse of the Renaissance, which revered in the literature, philosophy, and culture of classical antiquity a model that would never again be attainable. And so they preserved also the philological and interpretive methods that were supposed to explore the wisdom of antiquity and make it comprehensible. For many of these scholars, something was already true because a classical author had said it — and this applied in particular to Aristotle, who moreover had been taken into service for the Christian faith by Thomas Aquinas — although by no means uncritically. People read the Bible also, to a great extent, with that philological eye fixed purely on the wording. Quite in this spirit Lodovico delle Colombe wrote in late 1610 a treatise against the idea of the earth's movement, *Contro il moto della terra,* and circulated it in manuscript.[97] Galileo, too, had seen it personally, as proved by the marginal glosses in his handwriting on one copy.

Yet the astronomical, philosophical, and mathematical arguments of this treatise did not go down in history—aside from the fact that Galileo is said to have put a series of them in the mouth of the foolish Simplicio of his 1632 *Dialogo*. The course of events was defined rather by the last two pages of this work, on which delle Colombe draws an argument from the Bible against the movement of the earth:

> Weighty reasons against the Copernicans arise from Sacred Scripture; for in Psalm 103 it says: Thou hast founded the earth on its foundation, and when it says in Paralipomenon [1 Chronicles] 16: God founded the world [*orbem*] immovable, then *orbis,* as Abulensis [Alonso Tostado] remarks, should be understood to mean the earth. We see that the earth has its weight from the Book of Proverbs: The mountains were not yet set up with their heavy masses, and also from Isaiah: Who weighed the earth with its weight? Who held with three fingers the mass of the earth? And elsewhere in the Proverbs it says again: Stone is heavy and sand a burden. In the same passage we find the center of the universe assigned to the earth: The heaven is above, the earth here below.... Now if the earth belonged to a heavenly sphere, as Copernicus claims, then it would no longer be below, for the heaven is above. But the sun

[97] See Favaro, *Le Opere di G. Galilei,* vol. 3/1, 252–290. On Lodovico delle Colombe (1550–1635), see M. Muccillo, in *DBI* 38 (1990): 29–31.

> is not immovable either, for in the Book of Ecclesiastes we read: The sun rises and it goes down and returns again to its former place; by rising there once again, it travels across the meridian and inclines toward the west. Yet more! Did it not stop, in order to make victory possible for Joshua? Did it not go backwards at the time of King Hezekiah? Furthermore, we see from the Book of Genesis that the moon cannot be a second earth: God made two [celestial] lights, *i.e.* a greater and a smaller, and also the stars, so that they might shine upon the earth. The moon therefore cannot be a second earth; for if the earth, as our opponents claim, likewise served the moon as a light, as it does for us, then Sacred Scripture would be expressed inaccurately, since it speaks about only two lights and not about three. In it we never find the name "moon" or "light" applied to the earth, just as the name "earth" is never given to the moon. Moreover the moon is above, and therefore belongs to the dome of heaven, and not below, and therefore is not an earth.

"Perhaps the poor people," delle Colombe continues,

> have recourse to other, less literal interpretations of Sacred Scripture. But that will not work; for all theologians without exception teach that Sacred Scripture must be understood literally as much as possible and not in another sense; just think how often mystical interpretations bring all of philosophy and all of science into disorder! Melchior Canus, and with him all recent interpreters of the *Summa* by Saint Thomas, therefore laid down in his

> work *De locis theologicis* the thesis: Any interpreter of Sacred Scripture who proposes a doctrinal opinion contrary to the general understanding of the Church Fathers acts rashly. Moreover among theologians it is a general rule that a major philosophical error appears doubtful in theology, too, especially when, as is the case here, it has to do with a matter mentioned in Sacred Scripture. About the present question, Pineda says in his commentary on the Book of Job 9:6 that it can be traced back to Pythagoras, whatever fine names one may give to it nowadays. In order to avoid all suspicion of having exaggerated, I cite his own words: Some call this view a senseless, rash play on words that is dangerous to the faith; it originated with those ancient philosophers and was brought up again by Copernicus and Celio Calcagnini, more on account of their ingenuity than for any real benefit to philosophy or astronomy.... Our conclusions therefore are as follows: The earth is located in the center of the universe, and indeed immovably because of its weight. The sun revolves in its [fourth] heavenly sphere around the earth; the moon consists of parts that are more and less dense, but it is neither mountainous nor uneven; rather, as has been considered true until now, it is bounded by a smooth spherical surface.[98]

With that delle Colombe had answered the physical-astronomical question with the Bible. Presumably his intention in doing so was to

[98] German translation from Müller, *Galileo Galilei und das kopernikanische Weltsystem*, 81–83.

win over the theologians, too, as fellow combatants against Copernicus and Galileo. And in a society like the one then, the Bible and therefore the arguments derived from it had to have the highest authority, thus making unassailable the position that was defended in this way. It is significant, of course, that delle Colombe was probably aware of the hermeneutical problems that his plan raised, since he also referred to the necessity of an understanding of Sacred Scripture that is as literal as possible.

With the last two pages of his treatise, the Florentine opponent of Galileo had managed to induce the latter to start dealing now with questions of biblical interpretation. Of course they were by no means so foreign to Galileo as they would be to a natural scientist today, since education at that time included a wealth of religious and theological subjects, and it could be furthered and deepened by the eminent and zealous preaching that was conducted in the major centers such as Pisa, Padua, and Florence — Galileo's milieu. Then, too, Galileo had amicable dealings with numerous priests, religious, and even princes of the Church.

Galileo, who no doubt had recognized how dangerous delle Colombe's argumentation was, wrote to one of those princes, Cardinal Conti, asking him for information.[99] The query is lost, and only the answer has been preserved; Galileo's question seems to have been formulated as follows: "Is Sacred Scripture favorable to the Aristotelian teachings about the universe?" The cardinal replied on July 7, 1612, that this is certainly not the case with regard to the imperishability of the heaven, which Aristotle taught. Now as for a continual movement

[99] Carlo Conti (d. December 3, 1615) "ebbe indubbiamente una discreta cultura e una spiccata curiosità intellettuale che lo resero sensibile alle nuove scoperte galileiane" (no doubt had a fairly good education and an outstanding intellectual curiosity which made him responsive to the new discoveries by Galileo.) S. Andretta, in *DBI* 28 (1983): 376–378.

of the earth (not just turning on its axis), the cardinal opined that this is hardly contrary to Sacred Scripture. He cited for this view the Jesuit Lorin, who in his commentary on the Bible that had appeared in 1606 taught that nothing could be concluded from Ecclesiastes 1:4 against the movement of the earth that several ancient philosophers had maintained, nor did other passages from the Bible provide a strict proof against the earth's movement.[100] In this connection Lorin had spoken explicitly about Copernicus and Calcagnini. The cardinal himself thought that it is necessary to consider also that in the passages where the Bible speaks of the firmament and the like it uses everyday colloquial speech. However, the cardinal warned, on this subject one must go to work cautiously. Conti also referred to the Spanish theologian Diego Zúñiga, who maintained that the movement of the earth corresponds to Sacred Scripture rather than the contrary.[101] Yet, Conti did not think one could accept this opinion.

Probably Galileo pursued the problems of biblical interpretation further, especially in conversations with theological friends.

Then he received a letter from his friend, the Benedictine monk Count Benedetto Castelli, who was professor of mathematics in Pisa.[102] The letter was dated December 14, 1613, and reported about a social gathering involving Cosimo II, his wife, and his mother, in which Castelli too had participated. At it there had been an extensive and even heated discussion about Galileo's discoveries and Copernicus, and the Grand Duchess-Mother Christina had started to argue by citing the Bible. And people wanted Galileo to be aware of this.

[100] Jean de Lorin (1559–1634) had published, among other works, his *Commentarius in Ecclesiasten* in 1606 in Lyons. See B. Schneider, in *LThK2* 6 (1961): 1145.

[101] About Diego López de Zúñiga (Jacobus López Stunica, d. 1531 in Naples), see J. Schmid, in *LThK2* 9 (1964): 1125–1126.

[102] Benedetto Castelli (b. 1577/78, d. April 9, 1643) was descended from a family in Brescia. See A. Ferrari, in *DBI* 21 (1978): 686–690.

This was an echo of the question concerning "Copernicus and the Bible," which meanwhile was plainly being discussed far and wide.

But now Galileo surely had to speak up, if he wanted to do justice to his position at court. Since Galileo had received this veiled challenge through Castelli, the cardinal was therefore the right addressee for Galileo's well-known letter dated December 21, 1613, which was actually a treatise on the method of correct biblical interpretation. It covers eight pages in the edition of Galileo's works.[103]

The position that he adopts in this letter to Castelli is essentially the following: Sacred Scripture obviously cannot err, but its interpreters can, and in various ways. For instance, it would be a crude error to follow the literal meaning when Scripture speaks about God's hands, feet, or face, or about His wrath or repentance, or so forth. These anthropomorphisms require interpretation. This is true also for questions of natural science, for which we should rely on the Bible only in the very last instance, since the laws of nature operate with necessity, while the words of Scripture admit various meanings. Now since two truths cannot contradict each other, then in the case of an apparent contradiction between the Bible and definite findings of the natural sciences, theologians should make sure to bring their explanation of the Bible into harmony with science. Therefore it would be best to forbid fixing the explanation of ambiguous passages from the Bible to a particular meaning, the opposite of which might be proved someday by the natural sciences.

Moreover theologians should be content to interpret the truths of the faith that are necessary for the salvation of souls. God plainly intended to leave nature to the natural sciences. In particular, for example in astronomical questions, one should not argue from the Bible, which after all is liable to be misunderstood in many ways. The

[103] See Favaro, *Le Opere di G. Galilei*, 5:281–288; 71–77.

oft-cited passage in the book of Joshua, moreover, agrees less with Ptolemy than with Copernicus and his followers.

Galileo had this letter circulated in numerous copies, but he refrained from a print edition.

One echo of it was the oft-mentioned attack by the Dominican Tommaso Caccini from the pulpit of San Marco in Florence.[104] During the course of the usual continual interpretation of the Bible, he came on the Fourth Sunday of Advent, 1614, to speak about Joshua 10:12–14 and the "miracle of the sun" in the Aijalon Valley. This passage reads: "Then spoke Joshua to the LORD ...; and he said in the sight of Israel, 'Sun, stand thou still at Gibeon, and thou Moon in the valley of Aijalon.' And the sun stood still, and the moon stayed, until the nation took vengeance on their enemies.... The sun stayed in the midst of heaven, and did not hasten to go down for about a whole day." (RSVCE)

Even if we set aside the legends about this sermon that soon cropped up, the fact remains that Caccini insisted that the views of Galileo — without mentioning his name — contradicted the Bible. But with that he earned himself no laurel wreath. His own brother, living in Rome, wrote to him afterward — plainly the matter had been discussed there — an ironic, cutting letter in which he condemned the preacher's behavior, and the Dominican priest Maraffi, who held an influential position in the order, apologized to Galileo for his confrere, calling his action a *bestialità* (bit of foolishness) and distancing himself from it. Galileo himself weighed the option of filing a suit against Caccini in an ecclesiastical tribunal. When Prince Cesi vehemently advised him against it, pointing out the controlling influence of the Peripatetics, Galileo turned to his former student, Msgr. Piero Dini, who meanwhile had entered the service of the

[104] Tommaso Caccini, O.P. (1574–1648), was a gifted preacher who was much in demand, although he also had an ambitious and aggressive character. See P. Cristofolini, in *DBI* 16 (1973): 35–37.

Curia. He forwarded to him a copy of his letter to Castelli, knowing that it had caused a lively debate in Dominican circles. He asked Dini to show it to the Jesuit Fr. Christoph Grienberger, too, whom he called his *grandissimo amico et padrone* (very great friend and patron).[105] Cardinal Bellarmine was also to learn about the letter. In the conversation that Dini had with Bellarmine afterward, the latter already took up his position. He advised against forcing the matter, and opined that it would probably not come to a condemnation of Copernicus. At most one could formulate a few comments on the matter, to clarify that his cosmology is a mere hypothesis. The influential Grienberger shared this opinion, too, and Dini goes on to report that the Jesuit told him that he would have thought it better if Galileo had presented his astronomical proofs for Copernicus first and only then dealt with Sacred Scripture.

But now Galileo could no longer be dissuaded from pursuing the theological argument, although both Cesi and Dini emphatically advised him against it. Cesi, of course, forwarded to him now also the treatise by the Carmelite theologian Paolo Antonio Foscarini that had been printed shortly before in Naples, which presented the Copernican system as being compatible with the Bible.[106] The prince's commentary was almost enthusiastic: The treatise, which defended Copernicus without approaching Sacred Scripture, arrived at just the right moment. He did not think that Galileo's opponents would be provoked by it. The comprehensively educated, learned Msgr. Ciampoli reacted more skeptically; although he

[105] About Christoph Grienberger, S.J. (1561–1636), see F. Hammer, in *NDB* 7 (1966): 57.

[106] On this topic see Bruno Basile, "Galileo e il teologo 'Copernicano' Paolo Antonio Foscarini," *Rivista di Letteratura Italiana* 1 (1983): 63–96. Basile shows, among other things, that Galileo had no personal contact with Foscarini (1565–1616), although Galileo was familiar with his work and used it.

was convinced by Foscarini's arguments, he feared difficulties from the Holy Office — and he proved to be right about that.[107]

We should look at the substance of Foscarini's argument.[108]

After an attack against the intellectual rigidity with which the Peripatetics — he speaks only about "certain people" — held fast to the traditional opinions of ancient schools, he mentions among Galileo's other discoveries the phases of Venus as proof that this planet moves around the sun. He quotes the famous Simon Marius, who spoke about the necessity of a replacement for the complicated system of Ptolemy, and recommended the Copernican system for this purpose. Foscari attributes to this system plausibility at the very least and speaks furthermore about the possibility of a generally recognized proof for the thesis that Copernicus had accurately interpreted the actual cosmic situation. In this connection Foscarini now comes to speak also about the biblical argumentation against Copernicus.

He was convinced, he said, that only *apparent* contradictions could exist between Copernicus and Sacred Scripture; it is easy to resolve them by considering the manner of speaking in Sacred Scripture. In one case we must accept the figurative sense of a passage; in another case the Bible is expressed in everyday colloquial speech. Typical of the latter are all the biblical passages in which something human is predicated of God; for example, when it says that God walked in Paradise in the evening, became angry at man, and so on. When Scripture speaks about the earth standing firm, it means its continued existence in the midst of constant change, and when for instance it says about the apostle Paul that he was rapt to the third heaven, that by no means indicates an astronomical-cosmic quantity, but rather the dwelling place of the holy ones. Considerations of this

[107] About Giovanni Battista Ciampoli (1590–1643), see A. Ferrari, in *DBI* 25 (1981): 147–152.

[108] This summary of the content follows Paschini, *Vita e opere*, 303–310.

kind could show that the Copernican system is ultimately far more probable than Ptolemy's. Foscarini also made the important observation that the Church is infallible only in matters of faith and the salvation of souls, but can err in purely scientific questions.

Foscarini had not written in vain. Galileo read his treatise very carefully, and now in his great treatise, the Letter to the Grand Duchess-Mother Christina, when he set about presenting in greater detail the ideas about the problem of Copernicus and the Bible that he had already aired in his letters to Castelli and Dini, he made extensive use of the arguments formulated by Foscarini.

Moreover, for his views with regard to the interpretation of Scripture, Galileo cited Augustine and Jerome but also Thomas Aquinas. This was called for not only objectively but also strategically, for the Council of Trent had explicitly demanded that Sacred Scripture must be interpreted in harmony with the Church Fathers' understanding of the Bible. In this he could certainly rely on the groundwork done by his theological friends.

This Letter to Madama Christina Granduchessa di Toscana displays Galileo's understanding of science in a sort of summa. Therefore we will sketch here the essential lines of its contents.[109]

To begin with, Galileo laments the fact that his new discoveries have infuriated the representatives of the old doctrinal opinion against him, so that now they were taking the field with biblical arguments against him, although they themselves did not understand the Bible. Thus they were fighting against a system developed by a cleric, a cathedral rector, whose services a council had employed, whose work had been dedicated to a pope and was read throughout the

[109] See Favaro, *Le Opere di G. Galilei*, 5:309–348. For a critical summary with commentary, see Loretz, *Galilei und der Irrtum der Inquisition*, 72–113.

world without misgivings about his orthodoxy. But they were doing this in order to strike him, Galileo.

In the next section he attacks his opponents' misuse of the Bible by laying claim to it as an authority even in purely secular questions. Corpernicus, however, argued in a purely scientific manner and emphasized in his foreword to Paul III that Sacred Scripture can be used against him only if its meaning is twisted. He wrote, though, as a mathematician for mathematicians.

In impressive formulas that testify to his personal piety, Galileo then emphasizes his devout and reverent acknowledgment of the authority of the Bible and the Church's Magisterium.

Now, though, he attacks the set of problems that really motivate him. His opponents, he says, rejected Copernicus, reasoning that Sacred Scripture says that the earth stands firm, while the sun moves. Since Sacred Scripture can never err, they conclude now that any teaching that is not compatible with it must be rejected as false.

In contrast, Galileo does explicitly hold that the Bible is inerrant, but points out that its true, actual sense is often different from the mere literal sense and therefore is at first hidden. Otherwise we must attribute hands, feet, and a face to God, which obviously is heretical and blasphemous. If Scripture speaks nevertheless in this way, then it does so to adapt to colloquial speech and the crowd's ability to comprehend. But this is self-evident among theologians. Therefore, in questions of natural science one must argue first in a scientific way before one has recourse to scriptural authority. Moreover — and this is a central line of thought — one must reflect that nature and the Bible have their origin in the same way from God's Word. The Bible was dictated by the Holy Spirit; nature is the most obedient executor of God's will. But since the Bible in its manner of speaking must adapt to the general understanding of the people, while nature follows her laws unchangeably, therefore the findings of natural science are unchangeable, too, while the Bible requires interpretation.

Yet, according to the Church Father Tertullian, God reveals himself in nature no less than in Sacred Scripture. Furthermore God gave man an intellect and consequently wants him to use it, too, especially when it is a question of exploring nature. Besides, Sacred Scripture contains no astronomical theories, since the stars do not affect the salvation of human souls. A view that concerns the stars, in other words an object of the natural sciences, therefore cannot possibly be a heresy, that is, an erroneous religious doctrine. He now cites the late Cardinal Cesare Baronius (d. 1607), who was famous as the founder of the science of Church history, who said that the Holy Spirit does not intend to tell us through Sacred Scripture how the heavens go, but rather how we should go to Heaven.

Galileo substantiates this by a reference to Augustine and the Spanish exegete Benedict Pereira, S.J. whose work on Genesis had appeared in Rome in 1589–1598. Pereira characterizes very well how theology at that time viewed Galileo's question. Although personally convinced by Aristotle and his cosmology, he nevertheless formulates the important exegetical principle: If philosophers and mathematicians were to conclude, based on evidence, that there were not seven heavens, as has been taught hitherto, but rather eight or nine or more, then for a theologian and interpreter of Scripture it would be ignorance, if not foolishness, to reject and condemn their teaching as contrary or foreign to Sacred Scripture.[110]

Galileo is therefore firmly convinced of the compatibility of the two sources of knowledge — scriptural exegesis and natural science — although he expects theologians to take the trouble to reconsider, since biblical exegetes could err, even if the Bible cannot.

[110] See Grisar, *Galileistudien*, 258–259.

Now since new scientific discoveries about nature could be made every day, it would be very unwise to commit oneself to a particular opinion in these matters while citing the Bible. That would mean the end of all investigation in the natural sciences. But precisely in view of the millennial controversies, for example over the stability or movement of the earth and the sun, this investigation must remain open.

Now Galileo takes a further step by attacking those who tried to make theology the standard for all other sciences and thought that the findings of the natural sciences had to agree with the meaning of the Bible ascertained by theologians on the basis of the Church Fathers' understanding of Scripture. He correctly notes that this would mean demanding that natural scientists not see what they see, and not understand what they understand. There is, after all, a difference between a mathematician and a philosopher! By that he means, to use contemporary vocabulary, the difference between natural science, which works experimentally and empirically, and theoretical science which proceeds speculatively by drawing conclusions logically.

Again he cites Augustine with his challenge to theologians, in order to show that positively proven findings of the natural sciences do not contradict Sacred Scripture. Therefore, whatever is not positively proved and does contradict Scripture, Galileo continues, must be proved false and rejected. With that, of course, he shifted the burden of proof for matters of natural science onto the theologians.

From the discussion thus far he now draws the inference for the theological treatment of Copernicus. If an attempt were made to have his teaching condemned by the Church, one would have to forbid all astronomy and celestial observations—at a time when the truth was becoming increasingly evident. Even a partial prohibition of the work of Copernicus would result in great harm for the salvation of souls, since it would then be a sin to believe in a clearly known truth.

Then Galileo grapples with the view that says that we must understand all statements in the Bible concerning the things of nature in the literal sense, without further interpretation, when the unanimous exegesis of the Church Fathers understands it in this way. This however applies to the movement of the sun around the earth and to the fixed position of the latter, and therefore we are dealing here with a truth of the faith—to maintain the opposite would be heresy. Against this, Galileo now deploys the sure experimental knowledge of natural science—whereby admittedly he missed the point of the argument; namely, the binding character of the so-called *"consensus Patrum,"* the unanimous interpretation of Scripture by the Church Fathers.

Although we may hold fast to the literal meaning of the Bible wherever no knowledge gained empirically is available, when such knowledge is available, Galileo explains, then the facts should prompt us to search for the true sense of Scripture, which then no doubt is in conformity with the knowledge of natural science. For example, in view of the spherical form of the earth, Augustine did not interpret Psalm 104:2 literally—"you spread out the heavens like a tent"—and he gave theological reasons for proceeding in this way. And again Galileo makes the claim that biblical words concerning scientific objects should be interpreted according to the findings of natural science, but not vice versa, since the scientific knowledge of nature is also a gift of God. Besides Augustine he was able to cite for this argument the authority of St. Jerome and also St. Thomas Aquinas.

Besides, Galileo notes, as far as the patristic consensus is concerned, it does not exist at all on this question, since the movement of the earth and related matters were by no means among the questions discussed by the Church Fathers; nor did the councils deal with them. Therefore, if they left the question about cosmology open, then theology today can decide, but it should do so cautiously, as again, Augustine suggests. In short: the opponents of Copernicus, who dragged the Bible into this dispute, should be accused of nothing less than the misuse of Sacred Scripture. In addition, the pope himself cannot declare a thesis true or false without considering the nature of the matter, especially not if it is a dubious matter such as the movement of the earth.

In conclusion, Galileo repeats the interpretation of the passage from the book of Joshua that he had already advocated in writing to Castelli, which is much easier to understand according to the Copernican than according to the Ptolemaic system. Essentially, this letter to Madama Christina was the defense by a believer and a natural scientist of the independence of his investigation.[111]

[111] Why this treatise by Galileo was unable to convince his contemporaries was recently explained as follows by means of a textual analysis: "The ethos the author [Galileo] wished to project is undercut by his decision not to offer proof on the terms that were expected. The pathos he introduced because of his temperament led him to use appeals that must have rankled precisely those he needed to convince. Finally the ultimate test of the argument for his primary audience was in the logos, the scientific demonstrations he implied but did not present. Instead he carried his argument into the Court of his opponents the theologians, who, unfortunately, made the rules of the game." Jean Dietz Moss, "Galileo's Letter to Christina: Some Rhetorical Considerations," *Renaissance Quarterly* 36 (1983): 576. See also Olaf Pedersen, "Galileo and the Council of Trent: The Galileo Affair Revisited," *Journal for the History of Astronomy* 14 (1983): 1–29, who on the one hand points out the lacunae in the history of theology (e.g., Nicolas d'Oresme and Nikolaus von Kues, authorities in favor of Copernicus, are missing) but also writes,

> However, such unconscious shortcomings or conscious omissions are in the end of no great importance. The general outcome of the treatise is clear enough. In fact, Galileo succeeded admirably in stating a number of essential principles in such a way that they could be discussed apart from the technical deficiencies of his argumentation. Thus he made it clear that the Book of Nature and the Book of Scripture are complementary, but not contradictory, since they proceed from the same Author. Truth is one, and science is a legitimate way to truth independent of Revelation, although in a different sphere. Holy

Meanwhile, Foscarini had forwarded his treatise to Cardinal Bellarmine. The latter's reply, which was issued on April 12, 1615, was substantially a reply to Galileo's Letter to Madama Christina, also. Bellarmine praised Foscarini's study as ingenious and learned, yet advised both him and Galileo to advocate the Copernican teaching as a hypothesis, but not as agreeing with the cosmic reality. That, after all, would satisfy the needs of astronomy.[112] The claim about the stationary sun and the earth revolving around it, in contrast, is likely not only to challenge philosophers and theologians, but also to do harm to the faith itself, since it gives the impression that erroneous statements are being imputed to Sacred Scripture. Of course, Bellarmine notes, one can say that the problem in question is not a matter

---

> Scripture cannot be used against scientific statements once these are proved true beyond doubt by scientific methods. The language of the Bible is anthropomorphic and the literal sense does not always convey the true meaning. When difficulties arise one should not consult or abide with the Fathers on questions which they did not themselves discuss. And if the theologians wish to condemn a scientific statement it is their own task first to prove it false for scientific reasons. (ibid., 23)

Pedersen is incorrect when he accuses the Roman authorities of superficial, bureaucratic methods in their work and speaks about a nadir of Roman theology at the beginning of the seventeenth century. The name Bellarmine alone testifies to the contrary!

112 Paul K. Feyerabend remarks: "In modern terms astronomers may assert that one model has predictive advantages over another model, but they are mistaken when they assert that the model is, therefore, a faithful image of reality. This sensible principle is part and parcel of modern science." Feyerabend, "Galileo and the Tyranny of Truth," in *Galileo Affair,* 158. It is noteworthy that the seventeenth-century Scholastic already formulated a principle of the modern theory of science!

"Also many theories are developed as steps toward a more satisfactory, but as yet unknown view. They may be successful, but the very purpose for which they are introduced forbid us to draw realistic consequences from them" (ibid., 158). Thus the leading theorist of natural science in our day restores the view of the leading theologian of the seventeenth century.

of faith. While this is correct with regard to the topic, it still is a question of the contents of Sacred Scripture. However, if there were a real proof for the heliocentric system, one would have to proceed very carefully in interpreting Sacred Scripture and say, rather, that we had not understood its manner of speaking. At any rate, Bellarmine will not be convinced of the existence of such a proof until one is presented. After all, there is a big difference between saying that the Copernican system is consistent with all observations and maintaining that it is the only correct one. The latter appears to him more than doubtful, and as long as such doubts continue we must not abandon the previous interpretation of Sacred Scripture by the Church Fathers.[113]

This by no means official yet very personal and kind reply of the most important theologian of his time delineates the theoretical scientific problem exactly by allowing the Copernican system to stand as a hypothesis, as long as no scientifically cogent proof exists for its actual validity. Yet no ecclesiastical authority had ever forbidden or hindered efforts to produce that proof. What aroused the cardinal's misgivings was the problem of biblical interpretation. But even here Bellarmine left doors open; indeed he went so far as to agree with Galileo's view that in comparable questions biblical exegetes should be guided by the sure — but only the sure — results of investigation. Neither Copernicus nor Galileo, however, had been able to furnish proofs in the proper sense. One astronomer asks in this connection, "Why then does Galileo never adduce one of these new reasons? Why does he not study Kepler's

[113] See Jerome J. Langford, *Galileo, Science and the Church* (Ann Arbor: University of Michigan Press, 1971), 60–69; Feyerabend, *Der wissenschaftliche Realismus,* 88–89. Compare D'Addio, "Considerazioni," 35–36, who in addition presents important passages, either verbatim or in paraphrase, from an unpublished commentary on Thomas (1571) by Bellarmine.

Astronomia nova, the part which is properly speaking astronomical and scientific? Why does he never mention Kepler's gigantic advances in promoting the Copernican system?"[114] Why, we might add, does he not take a stand against his opponents' attempt to pursue astronomy with the Bible instead of with the telescope? Why does he let himself be distracted by an extraordinarily delicate theological dispute, for which he lacked competence?

In view of the overall situation, this was highly unwise, all the more so because Galileo had been told clearly what Cardinal del Monte had stated after a long conference with Cardinal Bellarmine and Galileo's two friends, Dini and Ciampoli: as long as Galileo deals with the Copernican system and its proofs, without getting into theological questions (*senza entrare nelle Scritture*), he will not experience the slightest difficulties. However, theologians would not just look on if he were to start giving biblical interpretations which, however ingenious they may be, departed from the general view of the Church Fathers. On February 18, 1615, Dini informed him that Cardinal Barberini—the future Urban VIII—had told him that there would be no more discussion of the Galileo case, and if the latter did not abandon the field of mathematics (*seguiterà di farlo come matematico*), he hoped that no one would cause him any trouble. These instructions acquire their full weight against the background of the events that had meanwhile taken place.[115]

The Florentine Dominicans, who probably were in close communication with the philosophers of the Aristotelian school in Pisa and like-minded circles in Florence, felt challenged by Galileo. Caccini traveled to Rome and there made contact with the leading

[114] Müller, *Galileo Galilei und das kopernikanische Weltsystem*, 106.

[115] However, D'Addio, "Considerazioni," correctly notes that Galileo was not the one who began with theological argumentation, but rather his opponents, whose arguments he now had to refute.

figures of his order and of the Curia. The matter had repercussions, and on March 19, 1615, Paul V ordered him to be interrogated about Galileo's views.[116] This was done the following day by the Commissioner of the Holy Office, Fr. Michelangelo Seghizzi, O.P., who also prepared a record of it. Caccini confirmed that Galileo taught that the earth moved, that he had a large following, and also that he was in correspondence with Paolo Sarpi. He himself in his sermon had pointed out the contradictions between Sacred Scripture and Galileo's teachings, and emphasized that Sacred Scripture must be interpreted according to the teaching of the Church Fathers, and should not be adapted to one's own views.[117] Furthermore several inquisitors were directed to investigate the spread of Galileo's teachings.[118] The result was the report filed with the Holy Office, the highest authority for the faith, that Galileo advocated the theses that the earth moved and the sun stood still, which — according to Caccini's understanding — contradicted Sacred Scripture, as interpreted by the Church Fathers. Other witnesses were heard, Galileo's work on sunspots was examined, and also the letter to Castelli, which was transmitted both to Fr. Grienberger and to Cardinal Bellarmine. Cardinal Barberini received a copy, too. The result of these investigations was the statement by a consultant of the Holy Office: Galileo now and then did employ an inappropriate manner of speaking, but did

[116] See S. M. Pagano and A. G. Luciani, eds., *I Documenti del processo di Galileo Galilei*, Pontificiae Academiae Scientiarum Scripta Varia 53 (Vatican City, 1984), 79.

[117] See ibid., 80–85. Paolo Sarpi (1552–1623) was a high-ranking superior in the Servite order, known for his anti-curial attitude. In the vehement dispute between Venice and Paul V about jurisdiction, he provided his native city with arguments against the pope, for which he was excommunicated. Galileo kept up a correspondence with him, by which he certainly did not recommend himself to the Curia. H. Jedin, in *LThK* 9:333–334.

[118] See Pagano and Luciani, *I Documenti del processo di Galileo Galilei*, 87–98.

not stray from the paths of Catholic doctrine. He said nothing about the Copernican question; he cared only about the theological problem of biblical exegesis.[119]

The quintessential message that Msgr. Dini derived from this was the advice to Galileo that it was time now to keep quiet and to work on solid reasons for the Copernican system — as far as the mathematical side of the question was concerned and also the biblical side.

That was the state of the affair, and Galileo would have done well to follow his friend's advice.

THE CONTROVERSY INTENSIFIES

The first clash with the Inquisition

But what prompted Galileo nevertheless to stir up the hornets' nest?

Was it because he was not content with a "philosophy of as-if" and wanted to get to the bottom of things themselves?[120] Maybe. Certainly, though, Galileo indulged in unjustified optimism, both with regard to how compelling his reasons were and also with respect to the willingness of his dialogue partners to acknowledge these reasons. A certain excess of self-assurance and the inclination to consider others as less intelligent than himself may likewise have prompted him. In any case, though, it was his enthusiasm, mightily fed by the telescope and the discoveries made with it, which caused him to assess incorrectly both the actual scientific state of affairs and also the theological-ecclesiastical situation.

In early December 1615, therefore, he arrived in Rome with a secretary and servants and took lodgings near Santa Trinità dei

[119] See Pagano and Luciani, *I Documenti del processo di Galileo Galilei,* 68–69. Unfortunately both the author and the date are unknown.

[120] See Hemleben, *Galileo Galilei in Selbstzeugnissen,* 89.

Monti on the Pincio Hill, in the house of the Florentine ambassador. The latter, who knew Galileo well, was not very edified by the visit, since he rightly feared complications. Immediately Galileo plunged directly into multifarious social contacts, which allowed him to get the impression of complete success. "I am very content; I see after all that the paths to the affirmation and propagation of my view are being smoothed, and I also feel that my health is improving,"[121] he wrote as early as December 12, 1615, to Chancellor Curzio Picchena in Florence.

Soon, however, the mood changed, and he complained bitterly about the intrigues of his opponents, the Dominicans and the Jesuits. Envy, ignorance, and godlessness, he claimed, were at work against him. A little self-criticism, though, would have kept Galileo from seeking the blame for it exclusively in other people. But one cannot overlook his inclination to assume envy, ill will, and enmity in others, and to feel that he was their bitterly persecuted victim: it is expressed in too many of Galileo's letters. In fact he had done much, everything possible, to provoke opposition. His appearance in the Roman salons, whose doors were wide open to the famous Florentine, was not very likely to win him friends. Msgr. Querenghi reported on this from Rome to Cardinal d'Este of Modena. The cardinal would be very pleased with Galileo, he wrote, if he could experience one of the numerous debates that the astronomer often started with fifteen or twenty people, who usually pressed him hard with questions but were then bested by his answers. He was so sure in discussion that he laughed them all to scorn. Although he hardly ever could convert someone to his new opinion, he nevertheless showed the inadequate cogency of the arguments used against him. The wittiest thing, in Querenghi's view, was Galileo's way of getting his opponents to

[121] Paschini, *Vita e opere*, 329.

skewer themselves, by first reaffirming their arguments, only to then ridicule them with irrefutable contrary reasons.[122]

And Galileo did this although he must have known about his opponents' efforts to move the Holy Office to intervene. They did for their cause nothing but what he was doing for his own: it was a matter of winning the approval of the highest ecclesiastical circles.

In this situation Galileo increasingly got himself into a predicament: the need for proof. What he had to offer thus far were well-founded doubts about Ptolemy and indications that pointed in the direction of Copernicus. He never deigned to grapple with Kepler and Tycho Brahe. Although he may have been convinced intuitively that the heliocentric system was correct, he provided not one single convincing proof in favor of it. Nevertheless, he now suddenly declared all those who still contradicted the Copernican system blockheads, intellectual pygmies who scarcely deserved to be called human beings.

In this rather hopeless situation that intuition and enthusiasm as well as vanity and obstinacy had maneuvered him into, he made the unsuccessful attempt to adduce the tides of the sea as a proof for the twofold movement of the earth. With that, of course, he was able to convince no one, even though he set it down in writing. But now the just-mentioned Cardinal Orsini, who was related to the Medici, brought up this theory in conversation in the circle of the cardinals gathered around Paul V as proof for Copernicus; he meant to support Galileo, but this must be viewed as the decisive misstep that provoked the intervention of the Holy Office.

Before that happened, on February 5, 1616, a visitor called on Galileo whom the latter had hardly expected: his Florentine opponent, the Dominican Tommaso Caccini. Two mutually

[122] See ibid., 330.

contradictory reports of this four-hour meeting exist, one from Galileo's pen, the other from the pen of Caccini's brother. Fra Tommaso apologized for his previous behavior toward Galileo and offered him any satisfaction that he wished. Since other additional visitors had arrived, they also talked about the question of the day. While Galileo reports that Caccini proved to be quite uninformed about it (which however did not prevent the others from agreeing with him), Caccini's brother said that Galileo was unable to answer Fra Tommaso's objections and therefore completely lost his temper. Be that as it may — both reports must be taken with a grain of salt — the conversation did not go satisfactorily, and its real purpose remains in the dark. Galileo himself saw in the Dominican's visit mere "hypocrisy, wickedness, deceit, and poisonous mania for persecution."[123] But Galileo routinely judged his opponents that way. It hardly occurred to him that they too might be prompted by the honorable motives which he self-evidently claimed for himself.

Meanwhile, it may have become clear to Galileo, too, that all these discussions were not so much about him personally as about the question of what position the Church should take with regard to Copernican cosmology. The role that he played in this was instead that of a catalyst which accelerated the process of developing opinion and helped to overcome the initial hesitation of the Holy Office to deal with the matter at all.[124]

Now, therefore, on February 23, 1616, two theses, which reproduced the essential content of Copernicus' teachings, were submitted to the consultors of the Holy Office for their

[123] Favaro, *Le Opere di G. Galilei,* 12:238. German translation from Müller, *Galileo Galilei und das kopernikanische Weltsystem,* 153–154.

[124] See Paschini, *Vita e opere,* 337–338.

appraisal.[125] They read: (1) "*Sol est centrum mundi et omnino immobilis motu locali,*" and (2) "*Terra non est centrum mundi, nec immobilis, sed secundum se totam movetur, etiam motu diurno.*"[126] In translation: (1) The sun is the center of the universe and does not move at all from one place to another, and (2) The earth is not the center of the universe, nor is it motionless; rather, it moves as a whole, day by day.

The next day a meeting of the consultors of the Holy Office was held, at which the aforesaid theses were discussed. Ten theologians participated, six of them Dominicans, among them the *Magister Sacri Palatii* (Master of the Sacred Palace, or Theologian of the Pontifical Household) and the Commissioner of the Holy Office. Presiding at the meeting was the archbishop of Armagh, Peter Lombard of Waterford. It resulted in the formulation of two statements, which then made up the judgment.

The first thesis about the motionlessness of the sun is foolish, philosophically nonsensical, and formally heretical, insofar as it explicitly contradicts both the wording and also the universally customary explanation of several passages of Sacred Scripture. With regard to the thesis about the earth's movement, they came to

[125] This is disputed by Fölsing, *Galileo Galilei — Prozess ohne Ende*, 335–338, who tries to prove that neither Copernicus nor Galileo had taught these theses — particularly with regard to the center of the universe. The theses are instead the result of a superficial comprehension of the matter and sloppy formulation by the consultors of the Holy Office. If that were so, then the only remarkable thing would be that Galileo never pointed out such a misunderstanding. He would necessarily feel also in that case that the condemnation did not apply to him at all — and this would put in an entirely new light his statements in the trial in 1633 that he had never considered something like that true, as well as his recantation. But this assumption would result in a plethora of absurdities that could scarcely be resolved. It therefore appears practically incomprehensible.

[126] Pagano and Luciani, *I Documenti del processo di Galileo Galilei*, 99.

the conclusion that it deserved the same rejection from the philosophical perspective, whereas viewed theologically it should be described as at least *in fide erronea*, "erroneous in faith."[127]

The fact that the consultants of the Holy Office passed a purely philosophical-theological judgment[128] shows that the natural scientific side of the problem and therefore natural scientific reasons for and against Copernicus were not discussed at all. Therefore, it was by no means, as is often claimed, fear of a collapse of the previous cosmology that moved the leading men of the Church. They were concerned instead about the inerrancy of Sacred Scripture, which they saw — albeit incorrectly — as being called into question by the claim of the Copernican system. In doing so, they disregarded the positive statements about Copernicus by Clement VII, Paul II, and Gregory XIV — Paul III had even accepted the dedication of the work by Copernicus — although the theologians were probably not aware of it at that moment.[129]

Although there is no documentary source for it, it can nevertheless be concluded from the further developments of the affair that this view was approved on the following day by the cardinals of the Holy Office at their meeting with the pope presiding.

Nevertheless, and surprisingly, from the judgment that the doctrine of the motionlessness of the sun is a heresy, they did not draw the only logical conclusion, namely the initiation of a heresy trial against Galileo. His name was not even mentioned and, astonishingly enough, his treatise about sunspots was not included in the

[127] See ibid., 99–100.

[128] Viewed from the perspective of Aristotelian metaphysics, this was certainly true. Even the creation of the earth could be described in those terms as foolish and absurd! See Stillman Drake, "Galileo and the Church," *Rivista di Studi Italiani* 1 (1983): 91–92.

[129] See D'Addio, "Considerazioni," 41.

censure, either, although in it he had indubitably taken up a position in favor of Copernicus.

What happened next probably can no longer be reconstructed unambiguously, although attempts to do so have been made repeatedly. We can select as our points of departure the two *"registrature"* (notes registered) in the acts of the trial dated February 25 and 26, 1616. According to the first one, Cardinal Mellini informed the members of the Holy Office that, on the basis of the aforesaid judgment by the theologians, the pope commissioned Cardinal Bellarmine to dissuade Galileo from his claims. If this was unsuccessful, then the Commissioner of the Holy Office was to issue to him before witnesses and in the presence of a notary the command neither to teach in the future along the lines of Copernicus nor to defend or discuss his teaching. If Galileo did not consent to this, he was to be taken into custody.[130]

The following *registratura* contains the report about the execution of the papal order, indeed in the form provided for the case in which Bellarmine had no success with Galileo. It also records the fact that Galileo had been summoned on February 26 by Bellarmine, who carried out his commission in the presence of Fr. Seghizzi. Next the Reverend Commissioner commanded Galileo in the name of the pope and of the Holy Office to stop teaching about the movement of the earth, and from now on neither to teach nor to defend it in word or writing; otherwise the Holy Office would initiate proceedings against him. Galileo promised to obey this command.[131]

[130] "Sanctissimus ordinavit cardinali Bellarmino, ut vocet coram se dictum Galileum eumque moneat ad deserendas [!] dictam opinionem; et si recusaverit parere, Pater Commissarius, coram notario et testibus, faciat illi praeceptum ut omnino abstineat huiusmodi doctrinam et opinionem docere aut defendere, seu de ea tractare; si vero non acquieverit, carceretur." Pagano and Luciani, *I Documenti del processo di Galileo Galilei*, 100–101.

[131] See ibid., 101–102. Two members of Bellarmine's household were witnesses.

A third document to be considered in this connection is an apology of Bellarmine for Galileo dated May 26, 1616, in which the cardinal confronted rumors that Galileo had been condemned or had had to recant and accept a penance.[132] It has been claimed repeatedly that there were internal contradictions between these documents, which led some to conclude that the *registratura* dated February 26 had been forged. A look at the original, which is preserved in the Vatican Archive and is accessible to any researcher, shows that the two entries were written by the same hand with the same ink, and thus should be described as simultaneous.[133]

A difficulty arises, however, since in the 1633 trial Galileo could no longer remember having learned about such a prohibition as the one noted in the *registratura* dated February 26, and promising to comply with it. Some suggest that we should assume that Galileo had allowed himself to be deceived about the seriousness of the prohibition by the urbane manner with which Cardinal Bellarmine dealt with the affair, which was painful for him as a learned man[134]; an obstacle to this explanation is the fact that it was not Bellarmine but rather the Dominican Seghizzi who was supposed to issue this command as Commissioner of the Holy Office.

Others have voiced the suspicion that one of Galileo's opponents in the domain of the Holy Office entered the *registratura* secretly into the file, so as to be able to set a trap for him later.[135] But if this were the case, then the one who made the entries would have betrayed himself, since both of them, the "genuine" one dated February 25 and the

132 See ibid., 138.

133 See Langford, *Galileo, Science and the Church*, 94–96. We should adhere to this view — even against Aubert, "L'état actuel," 153.

134 See Hans-Christian Freiesleben, *Der Prozess gegen Galilei*, Veröffentlichungen der Gesellschaft Hamburger Juristen 6 (Hamburg, no date [after 1965]), 14.

135 See Brodrick, *Robert Bellarmine*, 376–377.

"spurious" one dated February 26, are indubitably in the same handwriting, indeed are even written with the same ink. Therefore the notary himself or his scribe would be unmasked as a forger. Of course one would expect a perfect product from an insider-forger, which would not have shown the recognizable signs of a forgery, a point on which the later critics thought that they could support their hypothesis. Nevertheless, someone who is inclined to assume such a monstrous procedure as an "official falsification of the records" makes it known that he considers the Holy Office to be in fact a breeding ground of moral corruption from which such activity can be expected, without having to provide any proof for it. This, of course, would no longer be compatible with the claim to scientific seriousness.

Occasionally we read that Bellarmine's apology for Galileo dated May 26 contradicts the *registratura* dated February 26, but this too is based on a misunderstanding. The cardinal's concern could only have been to confront the rumors that were detrimental to Galileo's reputation, which is why he only had to say what had not happened — namely a condemnation, recantation, and penance. By no means, though, was an adequate presentation of the proceedings in late February to be expected in such a document.

A different and altogether plausible attempt to harmonize the sources starts from an opposition between the broadminded Jesuit Bellarmine, who was favorably disposed toward Galileo, and the Dominicans of the Inquisition who were hostile to Galileo. In fact Galileo himself in the year 1633 reports that Bellarmine had told him something that he intended to communicate only to the pope personally.[136] The assumption that he meant the explanation that Galileo should regard the specific prohibition by Fr. Seghizzi as not having been

[136] See the hearing on April 12, 1633, in Pagano and Luciani, *I Documenti del processo di Galileo Galilei*, 124–130.

issued, has more than a little probability in its favor.[137] Certainly this cannot be proved conclusively. But not everything that happens in such a situation, not every gesture or tone of voice, is reflected in the written sources on which the historian nevertheless must rely initially. Thus an unresolved remainder is no doubt left over.

But why should Galileo himself not be considered? The fact that in 1633 he could not remember receiving the prohibition expressed in the *registratura* dated February 26 of that year and pledging to abide by it, could very well be based in fact on an unconscious re-shaping of the actual proceeding in his recollection. Moreover, he never disputed the fact that he had received this prohibition, but only said that he could not remember it. To assume this is by no means to call into question Galileo's love for the truth. The more profoundly he was affected by these events, the more likely we can assume such a more or less conscious psychological process, the result of which came to light as a repression after the passage of sixteen years. In research on Luther, for example, it is a commonplace to speak about the reworking (*Umdichtung*) of the young Luther by the aging Reformer. Mario D'Addio, however, proposed another attempt to explain that is worth pondering. From the fact that the file contains no original, but only a text that is signed neither by the notary nor by witnesses, he concludes that the document therefore has no juridical significance whatsoever, that the prohibition was never issued. It was instead only a draft that was prepared for the eventuality.[138] The possibility that the entry could have been understood incorrectly almost twenty years later cannot be dismissed out of hand. The "falsification

[137] See Drake, *Galileo at Work*, 253ff.

[138] See D'Addio, "Considerazioni," 49–52. In this same context the likewise unfounded suspicion is voiced that the conversation had "appunto un carattere forse volutamente ambiguo" (precisely a perhaps deliberately ambiguous character). Vasoli, "Tradizione," 94.

of records," which has been claimed again and again since the 1860s has nevertheless become a constituent part of the legend, also of the *Leyenda negra* (Black Legend) of the Inquisition. In serious historiography, however, that song and dance ought to be over.[139]

During the session of the Holy Office on March 3, Cardinal Bellarmine informed the group that Galileo had bowed to the admonition to abandon his opinion about Copernicus in the future.

Now the only official decree from this phase of the trial was admittedly not issued by the Holy Office directly, but rather by the Congregation of the Index,[140] in other words the dicastery that was responsible for granting the imprimatur — ecclesiastical permission to print — or for censoring books. In stark contrast to what the consultants of the Holy Office decided, the decree by no means described the Copernican system as "heretical" but rather as "altogether opposed to Sacred Scripture." This deserves to be emphasized; for all the theological rejection, they nevertheless made the effort to spare Copernicus, and with him his followers, Galileo and the theologians Foscarini and Zúñiga, the accusation of heresy. The formulation that they found for this purpose — *sacrae scripturae omnino adversantem* (*i.e. doctrinam Copernicanam*) (the Copernican teaching, altogether opposed to Sacred Scripture) — had hitherto not been common in the terminology of the Holy Office and the Congregation of the Index.

We can hardly go wrong if we trace this obvious discrepancy between the appraisal of the consultors and the decree of the Congregation of the Index back to contrasts within the curial dicasteries between the followers of an intransigent Aristotelianism and those

139 See the relatively early remarks by Grisar, *Galileistudien*, 40–55.

140 See Pagano and Luciani, *I Documenti del processo di Galileo Galilei*, 102–103.

who were willing to make room for new findings. Plainly they also wanted to keep doors open for further development.

This prohibition, however, which affected Copernicus and the exegete Diego Zúñiga, was to remain in force not unconditionally but only until these works, which were explicitly described as valuable and beneficial, were corrected along the lines of the decree. An unrestricted prohibition applied only to Foscarini's booklet, whose publisher was thereupon taken into custody by Cardinal Carafa, the overzealous archbishop of Naples.[141]

A record from the archive of the Holy Office which was made available for the first time to the author, sheds new light on the proceedings following the sentence in February 1616.[142] At this stage the Congregation of the Index was occupied with the affair. Again Bellarmine was the one who at the pope's behest proposed to the cardinals of the Index for their deliberation the question of a prohibition of those works by Foscarini, Copernicus, and Zúñiga that maintained that the earth moves and the sun stands still. The session took place in Bellarmine's house, with Cardinals Bellarmine, Maffeo Barberini (the future Urban VIII), Bonifazio Caetani, Agostino Galamini, Orazio Lancellotti, Felice Centini, and the Theologian of the Pontifical Household present. After extensive discussion they decided that Foscarini's work should be forbidden, while with regard to Copernicus and Zúñiga they ruled that "*suspendendos esse donec corrigantur*" (they should be suspended until they were corrected). That meant, however, that their publications would be permitted after carrying out these corrections. It was also ordered that all other books "*idem docentes*" (that teach the same

[141] See Cardinal Carafa to the Holy Office, Naples, June 2, 1616, in ibid., 104.

[142] Walter Brandmüller and Egon Johannes Greipl, *Copernico, Galilei e la Chiesa: Fine della controversia (1820), Gli atti del Sant'Ufficio* (Florence: Olschki, 1992), 145–151.

thing) should be either forbidden or, like Copernicus and Zúñiga, "suspended."

The pope confirmed this decision and commanded the publication of it — subject to the interesting condition, however, that the aforementioned works were to be proclaimed forbidden or suspended not by themselves, but rather together with other books. The reason for this measure is not mentioned. Presumably they wanted to avoid the impression of a blow aimed exclusively at heliocentrism.

After a discussion between the Secretary of the Congregation of the Index and the Master of the Sacred Palace about which one of them should publish it, the pertinent decree was signed and made public; it was dated March 5.

With that, however, the affair was by no means concluded, for now the question was how the work by Copernicus could nevertheless be made accessible to the educated public, or how it could be obtained for them notwithstanding the sentences in February 1616.

The capable Msgr. Francesco Ingoli was commissioned to study this question and to draw up a recommended solution. He had already appeared in 1616 with a treatise against Galileo. Ingoli was the one who now commended the work by Copernicus as extremely useful for astronomy and much in demand by all, and now in a session of the Congregation of the Index on April 2 in Bellarmine's house he submitted his suggestion for proceeding to the prelates gathered there. In order to proceed safely, the cardinals decided to transmit Ingoli's exposé to the mathematicians at the Collegio Romano — in other words the Jesuits — for them to examine. Bellarmine made the arrangements, giving the task to Fathers Grienberger and Grassi, who after almost three months had prepared their unqualified positive vote, so that it could be submitted at the July 3 session of the Congregation of the Index. The congregation likewise followed Ingoli, and the corrections to the work by Copernicus were to be carried out according to his recommendations.[143]

After further sessions on February 28, 1619, and on January 31, 1620, which dealt with how the corrections should be published and publicized, the congregation decided on March 16, 1620, to issue the well-known decree, which was then supposed to appear on May 15, 1620.[144]

It is now established, as we showed, that Francesco Ingoli was its author. Therefore this decree listed the changes which, when carried out, would lift the prohibition of Copernicus's work. All of them aimed to preserve the character of the Copernican system as an astronomical working hypothesis. It is significant that the decree explicitly emphasized the scientific status of Copernicus and pointed out the great practical importance of his work—here they were probably thinking of the calendar reform. The list of a total of eleven passages that were to be corrected along the lines of the decree is appended to it.[145]

Therefore we cannot speak about a "death sentence passed on Copernicanism."[146] Like any other scholar, even Galileo could continue work on the "hypothesis," and everything that served to demonstrate the astronomical-mathematical appropriateness and the internal consistency of the Copernican system could promote the recognition that this was the adequate expression of the cosmic reality. He was prevented, however, from publishing such material by the special prohibition dated February 26, 1616.

[143] Ibid., 151–166, here in a synopsis from 1823.

[144] Ibid., 149. About the author, see Joseph Metzler, "Francesco Ingoli, der erste Sekretär der Kongregation (1578–1649)," in *Sacrae Congregationis de Propaganda Fide Memoria Rerum 1622–1972*, vol. 1/1 (Rome-Freiburg-Vienna, 1971), 197–232.

[145] Brandmüller and Greipl, *Copernico, Galilei e la Chiesa*, 149–151.

[146] Hemleben, *Galileo Galilei in Selbstzeugnissen*, 97.

On March 11 the pope himself had received Galileo for a conversation lasting three quarters of an hour, and as Galileo reports — exaggerating, perhaps — had assured him of his unshakable good will and had promised to protect him from his opponents as long as he lived. Galileo had made such a good impression at the Holy Office that it would not be easy to pay attention to new accusations against him.[147]

Nevertheless, exaggerated rumors about a reprimand of Galileo by the Inquisition quickly spread throughout Italy. In order to be able to counteract them, he asked Cardinal Bellarmine for his help, and the latter then issued on May 26, 1616, the apology already mentioned repeatedly. It read as follows:

> Since We, Robert Cardinal Bellarmine, have learned of repeated slanders about *Signore* Galileo Galilei, to the effect that he had to place into Our hands a recantation and that salutary works of penance were imposed on him on that occasion, We hereby declare upon request and in the interests of the truth that the aforesaid Galileo did not have to place a recantation of any opinion or teaching into Our hands or the hands of anyone else, neither here in Rome nor to Our knowledge in another place; nor were any penitential works whatsoever imposed on him. He was merely informed about the papal decision, published by the Congregation of the Index, that the doctrine attributed to Copernicus about the earth's movement around the sun and the central position of the stationary sun in the universe is contrary to Scripture — a doctrine that consequently may neither

[147] Favaro, *Le Opere di G. Galilei*, 12:248.

> be defended nor held. In witness whereof, We issued the present certificate in our own handwriting on this day, May 26, 1616.[148]

Nevertheless it must have been clear to Galileo that his great scientific plan to help Copernicus gain universal approval had run into an obstacle that would be difficult to overcome. No doubt he had to fear also a decrease in his prestige as a scholar and a weakening of his position at the court in Florence. This outcome of the affair was even worse for him because he plainly lacked the necessary scientific self-criticism and therefore was unshakably convinced that his stance was correct.

In debate with Grassi and Scheiner

In Florence they certainly viewed the situation less optimistically than Bellarmine did, for now a summons was issued to Galileo to return to Florence as soon as possible, in order to prevent further complications. Not in vain had the ambassador of the Medici in Rome, Piero Guicciardini, reported to his master on March 4 that Galileo, despite his urging, repeatedly tried to force his opinion on others, and that he was irascible, passionate, and downright fixated on this point. Thus, if he came into conflict with the Holy Office and the pope, he would plunge into very serious difficulties.[149] Two months later Guicciardini

[148] German translation from Müller, *Galileo Galilei und das kopernikanische Weltsystem*, 160. The text is in Pagano and Luciani, *I Documenti del processo di Galileo Galilei*,134–135 or 138 (according to the original). See furthermore on this subject *The Louvain Lectures (Lectiones Lovanienses) of Bellarmine and the Autograph Copy of his 1616 Declaration to Galileo*, Text in the Original Latin (Italian) with English Translation, Introduction, Commentary, and Notes by U. Baldini and G. V. Coyne, Vatican Observatory Publications, Studi Galileiani 1/2 (Vatican City, 1984).

[149] See Favaro, *Le Opere di G. Galilei*, 12:241–243.

put it even more clearly: Galileo displays here in Rome conduct that causes unusual annoyance. Everybody knows about the wild life that he and his companion Annibale Primi are leading. The ambassador also criticizes the extravagant lifestyle of his two guests and asks that Galileo be ordered back to Florence, so that even greater mischief might be avoided. However he also spoke of certain monks — probably Dominicans — and others whom Galileo had as opponents.

When the Florentine diplomat also mentioned that in Roman court society everyone strives to think in the same way as the prince (i.e., the pope), and if necessary at least pretends to think that way, he had thus cited another reason for the outcome of this first clash of Galileo with the Curia.

Accompanied by two warmly worded letters of recommendation by Cardinals Orsini and del Monte to the Grand Duke, Galileo set out in early June on his journey home to Florence.[150] Upon returning there, Galileo immediately resumed his studies. Shortly thereafter his daughter Virginia entered the convent in Arcetri as Sister Maria Celeste. The fact that this name was given to her may have been a veiled compliment to the much-admired father, whose "celestial" discoveries were well known even there. One year later her sister Livia followed Virginia's example, and they assigned to her the name Arcangela. That left their brother Vincenzio, whom Galileo legitimated quite late, namely in 1619, and thus made it possible for the youth to bear his name.

With interruptions due to illnesses and physical weakness, Galileo continued his observations of Jupiter, whereby he referred to the movements of Jupiter's moons in order to determine the difference between the geographical longitudes of various places, an idea that held promise for the future, although of course the attempt to apply

[150] See Müller, *Galileo Galilei und das kopernikanische Weltsystem*, 162–163.

it profitably in navigation failed at first because the technological prerequisites were lacking — a telescope that would stand still on board, precise clocks, and exact tables of the moon's movements. Nevertheless, via Cardinal Borja, Galileo immediately contacted the Spanish court, to which he tried to sell these findings in exchange for a yearly income of 6,000 ducats and a high distinction. In Madrid, however, they did not agree to the deal, although Galileo moderated his demands, so the affair came to nothing.[151]

The Florentine made a present of another invention to Archduke Leopold of Austria, his great admirer: a small telescope fastened to a sort of helmet, which could be used hands free. Along with the gift Galileo enclosed his writings about sunspots and about the tides. Of particular interest, however, is his accompanying letter dated May 23, 1618.[152] This letter shows the irony with which the writer viewed the Inquisition and its action in February 1616. He commended his two treatises, which defended the Copernican system, to the archduke "as a poem, or a reverie," to which he was attached, in the manner of a poet, and hence still felt a certain weakness for this dream. He also suggested to Leopold that the manuscripts be printed anonymously, and he did not fail to mention explicitly that he wished to secure for himself the priority rights to the "discoveries."

The year 1618 was one of his major, fateful years. It demonstrated the range of Galileo's character: we find him both on a pilgrimage to the Marian shrine in Loreto and also — once again — in a biting controversy. The appearance of three comets provided the occasion for it. The first of them was sighted in late August in the constellation Ursa Major; in mid-November the

[151] Se Galileo to the Florentine Ambassador in Madrid, Arcetri, June and December 25, 1617, in Favaro, *Le Opere di G. Galilei,* 12:358–361).

[152] See ibid., 389–392.

second one, with a long tail although not particularly bright, appeared in the sign of Hydra; and at the end of that month the third displayed a splendid trail from the Great Bear to Libra. However, while astronomers everywhere crowded around telescopes night after night, Galileo lay in bed, confined to it by physical weakness. Therefore, he could not participate on the basis of his own observations in the general discussion about the questions formulated by Kepler: "What are comets actually, where do they come from, what governs their movement, and in what way do they have something to say to the human race?" He was provoked to speak up nevertheless in this dispute by the Jesuit Orazio Grassi, who taught mathematics at the Roman College. In the auditorium of the college he had given a highly regarded public lecture in which he refuted the conventional theory, based on Aristotle, that comets could not have their origin and their place within the orbit of the moon. With that, Grassi, who explicitly cited Tycho Brahe, had adopted an altogether progressive position. The lecture appeared in print shortly thereafter.[153]

Had Galileo expected to be quoted by Grassi? Did he resent the Jesuit for dealing with astronomy? Be that as it may, he instructed his student Mario Guiducci to respond. Guiducci had been a student of the Collegio Romano, next had studied law in Pisa, and finally under Castelli's influence had studied the natural sciences with him and then also with Galileo. At the time he was thirty-four years old, and since 1607 he had been a member of the Accademia della Crusca in Florence. Now he

[153] For his lecture Grassi had not made exact observations, nor did he cite the pertinent results of his confrere Johann Baptist Cysat (Ingolstadt, 1619). See Juan Casanovas, "Il P. Orazio Grassi e le comete dell'anno 1618," in Galluzzi, *Novità celesti*, 307–313. About Oratio Grassi (1583–1654), see Carlos Sommervogel, *Bibliothèque de la Compagnie de Jésus*, vols. 1-IX (Bruxelles-Paris, 1890–1900; repr. Héverlé-Louvain, 1960), 1684–1686.

sat down at his writing desk to put into literary form the substance of the conversations that he had had with Galileo about comets. The decisive passages of the extant original, however, are written in Galileo's hand.

The result was a rather long lecture by Guiducci in the Academy, which soon afterward appeared in print. After resounding praise of the great Galileo, Guiducci went on the attack. First he took an incidental jab at the "plagiarist" Christoph Scheiner[154]; then he trained his sights on Grassi, whom he accused of having done nothing but subscribe to every word of Tycho Brahe. Soon, though, the professors of the Roman College in general were attacked, whose delusions were now to be exposed. However, the arguments that were now formulated — by Galileo, no less — do not stand firm even according to the epistemological criteria of the time. Grassi had come closer to the true state of affairs with his explanation than Galileo, for Tycho, whose reasoning he followed, had made the real breakthrough with regard to the study of comets. The judgment, "Even in error, Galileo thinks more scientifically than his opponents,"[155] is rhetorically very effective, perhaps, but objectively quite dubious.

Apart from its scientific substance, "Guiducci's" treatise (there could be no doubt about its real author) was an insulting attack on those Jesuit colleagues whose esteem for Galileo had been largely untarnished even after his attacks on Scheiner, and a sign of monumental ingratitude to the Roman College, which had prepared genuine triumphal processions for Galileo during his visit to Rome less than ten years earlier. Guiducci, as a former student of the Collegio Romano, was particularly embarrassed by the situation and tried to apologize to Grassi, who subsequently continued to show him such great generosity and kindness that Guiducci felt ashamed.

[154] About Christoph Scheiner (1575–1650), see Sommervogel, *Bibliothèque de la Compagnie de Jésus,* 734–740.

[155] According to Hemleben, *Galileo Galilei in Selbstzeugnissen,* 103.

Nevertheless, for the sake of the scientific prestige of the Roman College and of his order, Grassi owed a response to Galileo. He delivered it under the easily decoded pseudonym Lotario Sarsi Sigensano, ostensibly a student of Grassi. Recently the well-founded suspicion has been voiced that Scheiner made his arguments against Galileo's or Guiducci's theories about comets available to Grassi and urged him to include them in his work. Grassi, a sociable, likable character, then felt obliged to assure Galileo through friends of his undiminished admiration.[156] We could see an indication of this in the fact that Scheiner published nothing of the sort under his own name.

In his *Libra astronomica ac philosophica* (1624) Grassi elegantly and objectively comes to grips with Galileo's main arguments, but here and there still has polite praise for his opponent as well. "The common judgment that the *Libra* contains strong attacks against Galileo or ridicule cannot possibly originate from a firsthand reading of it; it is only an echo of what Galileo's overzealous and flattering students and friends write to him about his opponent."[157]

When Galileo got to see this treatise, it caused him outbursts of his choleric temperament, as the still extant glosses that he wrote in the margins testify. "Ignorant," "pedant," "malicious fool," "colossal blockhead" are the epithets that he assigns to Grassi, whom he also calls the biggest ox that he had ever seen, a liar, and a betrayer.[158]

Apparently, though, Grassi's arguments left Galileo in a predicament, because it took three years for his response to appear in print. The urgent warnings of his best friends could not dissuade him from publishing it. Finally, the Lincei — members of the "Academy of the Lynx" — recommended that Galileo should reply to Grassi in the

[156] See Drake, "Galileo and the Church," 276.

[157] Grisar, *Galileistudien*, 325. Along the same lines, see Drake, "Galileo and the Church," 277.

[158] See Müller, *Der Galilei-Prozess*, 18–19.

form of a letter to one of their members, Virginio Cesarini.[159] After Cesi and Ciampoli had read and corrected the manuscript, it was submitted to the official censor, the Dominican priest Riccardi, who granted the imprimatur in extremely flattering words which naturally were placed at the front of the book.

The title reads *Il Saggiatore*—The Assayer, or Goldsmith's Scale—"on which the contents of the *Libra astronomica* ... by Lotario Sarsi Sigensano are weighed."

In Galileo scholarship from the turn of the twentieth century the scientific value of this treatise was very highly esteemed. Some for example spoke about an inimitable model of scientific polemics.[160] Moreover, it was viewed as a pearl of Italian literature. In literature about the history of science around the same period, in contrast, it is regarded as scarcely worth mentioning.

Recently, however, the *Saggiatore* has been thrust into the limelight again. A surprising thesis that contradicts all previous findings is now attached to the book: Galileo had only *apparently* been condemned on account of his Copernican teachings. What they really charged him with was his atomistic-nominalistic natural philosophy, which the Holy Office feared might endanger the dogma of eucharistic transubstantiation. The basis for this claim was a document found in the archive of the Holy Office containing a discussion of the *Saggiatore*. The fact that this paper shows no sender nor addressee nor date nor context and contains no complete text either, urges caution; however, that in itself, is not enough to cast doubt on this thesis. The decisive fact, nevertheless, is that, quite unlike Galileo's *Dialogo* (which will be discussed soon), his *Saggiatore*, which is said to have contained these very heresies, was not put on the Index but remained

159 About Virginio Cesarini (1595–1624), see C. Mutini, in *DBI* 24 (1980): 198–201.

160 See Müller, *Der Galilei-Prozess*, 25.

quite unobjectionable. This pulls the rug out completely from under this somewhat fantastic assumption.[161]

In fact, the mere structure of the work is not very apt to arouse interest: Galileo proceeds by quoting Grassi's individual theses and then picking them apart — fifty-three in all. The tone is as polemical as possible; after all, Galileo accuses his opponent of malicious intent, jealous intrigue, and outrageous judgment. He calls Grassi a poisonous scorpion that he will tread on and destroy in its own venom. He vented all his ill humor once again when Grassi composed a voluminous response, to which Galileo, however, no longer replied. He contented himself with making marginal notes in Grassi's book: for example, "You ass, buffalo, common sluggard, blockhead, miserable forger, vulgar fellow, liar, dumb cattle, thick-headed lout" — an impressive florilegium![162]

It was an altogether unpleasant and ultimately unproductive controversy, which on both sides was more personal than scientific.

With that, things had reached the point where relations between Galileo and the Jesuits of the Roman College were seriously disturbed from then on. Even though we certainly cannot speak about enmity toward Galileo on the part of Grienberger and Grassi, nevertheless the former by his attacks had helped strengthen the solidarity of the members of the order with Fathers Scheiner and Grassi. Thus their earlier admiration turned into aversion, which, given the influence of the Jesuits, necessarily had its consequences.

[161] See Pietro Redondi, *Galileo Heretic* (Princeton, N.J.: Princeton University Press, 1987). Besides very interesting remarks about intellectual history and the history of science, the book contains many conjectures and is not without its romantic features. On this subject see, among others, the incisive critique by V. Ferrone and M. Firpo, "Galilei tra Inquisitori e Microstorici," *Rivista Storica Italiana* 97 (1985): 177–238. Also Redondi's response and the reply by Ferrone and Firpo, ibid., 934–956 and 957–968.

[162] See Müller, *Der Galilei-Prozess*, 39.

Nevertheless, Galileo's student Guiducci, who had remained in Rome and fallen sick there, was able to report friendly visits from Grassi. The latter even showed that he was quite open to the assumption that the earth moves, and he declared, exactly along the lines of Bellarmine's position, that when irrefutable proofs for this were presented, Sacred Scripture would just have to be interpreted differently.[163]

The situation was further aggravated by Galileo's attacks on the astronomer Scheiner, whom he accused by Guiducci's pen of claiming Galileo's discoveries for himself. Galileo himself then repeated the accusation in his *Saggiatore,* without mentioning Scheiner's name, however. The book had appeared just as Scheiner had arrived in Rome for a stay lasting several years. In fact sunspots were discovered by Scheiner, Galileo, Thomas Harriot, and Johannes around the year 1611, almost simultaneously and independently of one another. But while Galileo accurately described them as being located on the sun itself, Scheiner, so as not to contradict the opinion of the Aristotelian school about the sun's purity and serenity, declared them to be small bodies revolving around the sun a short distance away. On the other hand, Galileo no doubt was wrong in the quarrel over priority in discovering the phenomenon. Then too, Scheiner's discoveries go farther on several points than Galileo's do. For example, he was the first to define very exactly the position of the sun's equator and the rotating elements of the sun, and he also recognized that sunspots have a movement of their own which has nothing to do with the sun's turning on its axis, and several more points.[164] As in the quarrel over priority concerning the proportional compass, so too in this dispute, Galileo gave free rein to his bilious temperament and thus proved to

[163] See Drake, "Galileo and the Church," 289–293.

[164] See Hans-Christian Freiesleben, *Galilei als Forscher* (Darmstadt: Wissenschaftliche Buchgesellschaft, 1968), 68–69.

be unfair and arrogant, while Scheiner in his great work about the sun, *Rosa Ursina*, which would appear later in 1630, dealt roughly with Galileo's claims and was also harsh in tone, yet without falling into personal polemics. How good it would have been, though, for Galileo to have Scheiner too for a friend, as he once had his confreres Clavius and Grienberger!

Now, between the composition and the appearance of the *Saggiatore* there had been a change of pontificate, which was of great significance for Galileo. On August 6, 1623, after a difficult conclave that had lasted since July 19, Maffeo Barberini ascended the See of Peter as Urban VIII. As a cardinal he had already been an admirer of Galileo and had celebrated his discovery of the sunspots in a Latin ode. The pope called two figures from Galileo's closest circle of friends to be in his innermost entourage: the Linceo (member of the "Academy of the Lynx") Virginio Cesarini became *camerlengo* (chief chamberlain), while Ciampoli remained secretary for correspondence with the princes and became the private chamberlain on duty, and Prince Cesi enjoyed special papal benevolence.

Borne up in that hour on the wings of favor, they decided to dedicate Galileo's *Saggiatore* to the pope. He accepted the dedication and read the book, at least the more important passages of it. Circumstances were now so favorable that Galileo decided to set out for Rome as soon as his health allowed. After a two-week stay with Cesi at his castle in Acquasparta, he arrived in Rome on April 23, 1624. He met with a reception that caused the experiences of the previous eight years to pale. Again the doors opened to him. He sat at table with cardinals and there had an opportunity for numerous encounters with the experts and officials of the Curia.

In particular Cardinal Eitel Friedrich von Hohenzollern-Sigmaringen also interceded with the pope for him. Urban presented Galileo with a picture, a gold medal, and a silver medal. More important than this was a brief to Grand Duke Ferdinand of Tuscany drawn up by Ciampoli, full of praise and good will for Galileo.

These signs of papal favor, however, should not have deceived him about the true situation with regard to the cosmological question that most deeply motivated him. Certainly, the pope admired the Florentine expert. In an audience he had even made a significant remark to Cardinal von Hohenzollern when the latter — at Galileo's request — had appealed to him on behalf of the heliocentric cosmology. The pope had replied that the teaching of Copernicus had never been condemned as heretical, which could not happen, but was only described as a *sententia temeraria* — "a rash opinion" — yet there was no reason to fear that someone would prove it correct. Galileo's other theological conversation partners, the Dominican Niccolò Riccardi and the German convert Caspar Schoppe, were of the opinion that this was not a question of faith that was supposed to be answered by consulting Sacred Scripture.[165]

All these communications show how much Galileo was dominated, now as before, by the idea of helping Copernicus gain approval — so much so that he greatly overestimated the opportunities to do this that were offered to him now since Urban's election. They were limited by the distinction between knowledge that rises to the level of certainty and a mere working hypothesis. Even Urban VIII could not decide to overstep this boundary.

Galileo, however, would not let go of the question about cosmology. Now he even set out to respond to the treatise written by Msgr. Francesco Ingoli against Copernicus as early as 1616, before the Congregation of the Index issued its decree. Why it took him eight years to do this is no longer entirely clear. Did he consider

[165] See Galileo to Cesi, Rome, June 8, 1624, in Favaro, *Le Opere di G. Galilei,* 13:182–183.

Ingoli's arguments out of the question, or did he intend to respect the prohibition imposed on him by the Inquisition? Each of these reasons may have affected his decision. Ingoli, the theologian, who since the foundation of the Congregation for the Propagation of the Faith by Gregory XV in the year 1622 had served as its first and extraordinarily successful secretary, had systematically compiled in his treatise the common arguments against Copernicus, dividing them into mathematical, philosophical, and theological reasons.[166] At the conclusion of his work, he had challenged Galileo to resolve these difficulties, with the exception of the theological ones. Since Ingoli was not a specialist in astronomy, this must have been very easy for Galileo. In fact his answers clearly show the superiority of their author. However, it is plainly evident from Galileo's remarks that they are based in detail on Kepler's *Epitome Astronomiae Copernicanae* which appeared in 1618, although Galileo does not cite Kepler.[167] Once again the manuscript was forwarded to his friends in Rome, who were supposed to smooth the path for the booklet. They, however, headed by Cesi, refused to publish it, since Galileo would thereby publicly violate the 1616 decree of the Congregation of the Index. They advised him to let the matter rest — time would tell.

The "Dialogo sopra i due massimi sistemi del mondo"

Meanwhile, however, Galileo was already busy with his great work about cosmology, though he spoke again and again about his *Dialogo*, which he was preoccupied with writing. In the *Sidereus Nuncius* he had already spoken repeatedly about this plan, which no doubt had been inspired by the experience of his telescopic discoveries.

[166] Ingoli's treatise is in Favaro, *Le Opere di G. Galilei,* 5:403–412.

[167] See Müller, *Der Galilei-Prozess,* 48–57.

To all appearances, the purpose that he had in mind was to provide the decisive proof for the validity of the Copernican system. No doubt he had had substantial success on the way to that goal, inasmuch as he had been able to invalidate a series of arguments previously adduced against Copernicus. Ultimately, however, this did no more than clear away obstacles. Only very gradually was he approaching the goal itself. In particular he had to comply with the publicly known decree of the Congregation of the Index and the prohibition imposed on him personally in 1616, if he wanted to avoid another conflict with the Curia.

Even though the ascent of Urban VIII to the papal throne had changed the situation in Galileo's favor, the fact that it seemed even to Galileo's friends unwise to publish his response to Ingoli showed how necessary caution and restraint still were.

Nevertheless Galileo, urged repeatedly by his friends to do so, remained at work. What mattered to him was the broadest possible audience; this is evident in his choice of neither the scientifically strict form of a treatise nor of the Latin language usually employed for such purposes (which would also have guaranteed him a learned readership outside of Italy). Instead he wrote in the Italian language and chose the literary form of a dialogue—actually a trialogue—which allowed him a looser train of thought as well as personal attacks and other digressions. People were more likely and more eager to read something like that than an astronomical treatise in Latin.

This dialogue form had another advantage, too: the author could hide behind each of the speaking persons, and therefore to some extent his own opinion could not be determined.

The dialogue is divided into four rounds of conversation, one per day. Of the fictional participants, two bear the names of deceased friends of Galileo from his days in Padua, Filippo Salviati and Sagredo. The former, who died young, had been an especially gifted student of Galileo. As a rule he also allows him to express his own opinion and to determine the course of the conversation. To Sagredo, in contrast, who was neither an astronomer nor a mathematician, he assigns the role of the one who advances the dialogue with his astute and sympathetic questions. The real adversary is the limited Simplicio, on whose lips are placed the antiquated arguments of the Peripatetic school. His name has a sophisticated double meaning: first, there was a famous commentator on Aristotle by this name, but also the Aristotelian was supposed to be viewed ironically as simple-minded. Now when he, of all the characters, is made the mouthpiece of ideas that had been expressed by Jesuits at the Roman College or even by the pope himself, the covertly expressed mockery could hardly go unnoticed. In particular, though, Galileo made fun of an argument taken from Thomas Aquinas's commentary on Aristotle and used by Urban VIII, which said that the explanations of the astronomers for the movements of the heavenly bodies would not necessarily have to be true also, for it would be quite possible that they would have to be explained in a completely different way which until now had remained unknown to men.[168] It betrays at least unjustified epistemological optimism when Galileo lightly dismisses such considerations.

In order to understand the events following the publication of the *Dialogo*, it is helpful to reprint here verbatim excerpts from Galileo's foreword:

> Several years ago [in 1616] there was published in Rome a salutary edict which, in order to obviate the

[168] See "Commentaria in libros Aristotelis de caelo et mundo," Lib. II, Lect. XVII, in *S. Thomae Aquinatis Opera omnia, ed. iussu impensaque Leonis XIII. P. M.*, III (Rome, 1886), 186–189.

dangerous tendencies of our present age, imposed a seasonable silence upon the Pythagorean opinion that the earth moves. There were those who impudently asserted that this decree had its origin not in judicious inquiry, but in passion none too well informed. Complaints were to be heard that advisers who were totally unskilled at astronomical observations ought not to clip the wings of reflective intellects by means of rash prohibitions.

Upon hearing such carping insolence, my zeal could not be contained. Being thoroughly informed about that prudent determination, I decided to appear openly in the theater of the world as a witness of the sober truth. I was at that time in Rome; I was not only received by the most eminent prelates of that Court, but had their applause; indeed this decree was not published without some previous notice of it having been given to me. Therefore I propose in the present work to show to foreign nations that as much is understood of this matter in Italy, and particularly in Rome, as transalpine diligence can ever have imagined. Collecting all the reflections that properly concern the Copernican system, I shall make it known that everything was brought before the attention of the Roman censorship, and that there proceed from this clime not only dogmas for the welfare of the soul, but ingenious discoveries for the delight of the mind as well.

To this end I have taken the Copernican side in the discourse, proceeding as with a pure mathematical hypothesis and striving by every artifice to represent it as superior to supposing the earth motionless — not, indeed absolutely, but as against the arguments of some professed Peripatetics. These men indeed deserve not even that name, for they do not walk about; they are content to adore the shadows, philosophizing not with due circumspection but merely from having memorized a few ill-understood principles.

Three principal headings are treated. First, I shall try to show that all experiments practicable upon the earth are insufficient measures for proving its mobility, since they are indifferently adaptable to an earth in motion or at rest. I hope in so doing to reveal many observations unknown to the ancients.

Secondly, the celestial phenomena will be examined, strengthening the Copernican hypothesis until it might seem that this must triumph absolutely. Here new reflections are adjoined which might be used in order to simplify astronomy, though not because of any necessity imported by nature.

In the third place, I shall propose an ingenious speculation. It happens that long ago [in 1616] I said that the unsolved problem of the ocean tides might receive some light from assuming the motion of the earth. This assertion of mine, passing by word of mouth, found loving fathers who adopted it as a child of their own ingenuity. Now, so that no stranger may ever appear, arming himself with our weapons, to charge us with want of attention to such an important matter, I have thought it good to reveal those probabilities which might render this plausible, given that the earth moves. I hope that from these considerations the world will come to know that although other nations have navigated more, we have not theorized less. It is not from failing to take count of what others have thought that we have yielded to asserting that the earth is motionless, and holding the contrary to be a mere mathematical caprice, but (if for nothing else) for those reasons that are supplied by piety, religion, the knowledge of Divine Omnipotence, and a consciousness of the limitations of the human mind.

I have thought it most appropriate to explain these concepts in the form of (scientific) dialogues, which, not being restricted to the rigorous observance of mathematical laws, make room also for digressions which are sometimes no less interesting than the principal argument.[169]

[169] Favaro, *Le Opere di G. Galilei,* 7:27; translated from Italian by Stillman Drake. Also, for an interpretation which is not unreservedly satisfactory but nonetheless is characteristic of the current view of the problem, see Angiolo Procissi, "Commento alla prefazione 'Al discreto lettore' premessa da Galileo al 'Dialogo sopra i due massimi sistemi del mondo,'" *Atti e memorie dell'Accademia Toscana di Scienze e Lettere "La Colombaria"* 42 (1977): 95–120. See also Stillman Drake, "Galileo's 'Dialogue': Al discreto lettore," *Scientia* 117 (1982): 249–261, whose interpretation — that the common opinion that Galileo intended to dupe the Inquisition by his foreword to the *Dialogo* — is too stupid to warrant a refutation, can hardly be right! The theses presented here would generally need extensive discussion, which however is not possible here.

With that, however, Galileo trapped himself in his own fine-spun net. What did he take his readers for? Did he really think that they would let themselves be deceived by him?

The introduction already contains a blatant untruth, when he pretends that his book was written only because he wants to show non-Catholics that the explanation for the Roman prohibition of Copernicus is by no means a lack of knowledge about the state of the problem, but rather that the authorities in Rome are well informed about the arguments pro and contra. When he goes on to write that in recent years a salutary decree was issued in Rome which in a timely manner commanded silence about the Pythagorean teaching about the earth's movement, this was bold-faced mockery. Indeed, the text of the book left not even the slightest doubt that Galileo was convinced of the opposite: Salviati always has the stronger arguments, and Simplicio's reasons are ineffectual, if not outright laughable. Nevertheless Galileo feigns agreement and compliance with the 1616 decree of the Congregation of the Index. Anyone able to read, however, would have to wonder here whether, given Salviati's proofs, which are presented as compelling, the decree may be based on error after all, and not on Copernicus.

But now in Rome, too, there were people who understood something about astronomy, mathematics, and physics. And it must have been striking to those very same people, if they suppressed their annoyance and read Galileo's book soberly, that even this time he was unable to produce any arguments in favor of Copernicus except the ones he had already presented in 1616. Nor did he satisfy the demand that Bellarmine had stipulated at that time, that one must have compelling proofs for Copernicus before one can decide to interpret the Bible passages that are regarded as pertinent in any other way than literally and thus resolve the contradiction between Copernicus and the Bible. In fact Galileo presented no proof. Not one single argument of his can be regarded as such, least of all his theory about the tides. Nevertheless — if we disregard the last-mentioned one — taken together, the arguments were not without some persuasive force for anyone who had bid farewell to the purely deductive, Peripatetic way of thinking and was willing to gain knowledge from observations and the new celestial mechanics, as Galileo taught.[170]

While no expert in the history of the natural sciences disputes this fact concerning Galileo's lack of proof, it is concealed again and again in presentations of "the Galileo case." Nevertheless it makes a considerable difference whether the Church opposed knowledge that had been proved beyond all doubt or merely an assumption that had not been finally proved. And that was precisely the question, both in 1616 and also concerning Galileo's *Dialogo*, which now in early 1630 had at last been finished and was supposed to be sent to the printer. Frequent illness had prevented the aging Galileo from working continually.

The publication was supposed to take place in Rome, where influential friends were at work who could remove the obstacles. This group meanwhile included Castelli also, who had been appointed professor of mathematics at the Pontifical Sapienza University, and the Dominican Riccardi, since 1629 the Theologian of the Pontifical Household.[171] In this capacity Riccardi was responsible also for granting the imprimatur for all books appearing in Rome. It is interesting that Riccardi was a cousin of the wife of the ambassador from Tuscany in Rome. When Ciampoli made it known through Castelli

[170] See Filippo Soccorsi, "Il Processo di Galileo," in *Miscellanea Galileiana* 3:900–912.

[171] See Ambrosius K. Eszer, "Niccolò Riccardi, OP, il 'Padre Mostro' (1585–1639)," *Angelicum* 60 (1983): 428–461, who with good reasons protects Riccardi against all sorts of criticism.

that Galileo's personal presence in Rome would help to overcome all existing obstacles, Galileo set out once again for Rome, where he arrived after a two-day journey on May 3, 1630. This time, too, he took up lodgings in the Florentine embassy on the Pincio.

Negotiations concerning permission to print the *Dialogo* began immediately.[172] The prerequisites were more favorable than ever; after all, the pope himself had told Tommaso Campanella that the 1616 Decree would never have been issued if it had depended on him.[173] Urban VIII received Galileo on May 18 in a long audience during which no doubt they discussed the *Dialogo* and its contents. The benevolence that the pope showed to Galileo again this time was expressed in his promise of a pension for the scholar.

But even this renewed sign of the pope's favor in no way changed the fact that even now they could speak only about a hypothetical acceptance of the Copernican theory. Fr. Riccardi was fully aware of this. But since he himself was not a specialist, he passed the manuscript of the *Dialogo* on to a Dominican confrere, the mathematician Fr. Visconti, whose opinion was that permission to print should be granted if several formulas were changed to suggest the hypothetical validity of the heliocentric system. Fr. Visconti himself was the one who undertook to make these changes together with Galileo. When he submitted the results to Fr. Riccardi, the latter was satisfied and announced that on the following day, June 17, he would consult again with the pope — only a few insignificant details stood in the way of an imprimatur. Nevertheless, the affair dragged on, and since Galileo wanted to escape the Roman heat, Riccardi did something that he should not have done. He gave Galileo a blank signature, on the condition that

[172] Fr. Riccardi reports about them in detail in late May 1631. See Pagano and Luciani, *I Documenti del processo di Galileo Galilei*, 105–108.

[173] About Tommaso Campanella, O.P. (1568–1639), see L. Firpo, in *DBI* 17 (1974): 372–401.

Galileo himself would make a few more changes and then send the manuscript to Rome, where Prince Cesi, for instance, would supervise the printing. However, the galley proofs would have to be submitted once again to Riccardi before printing began.

Now the prince died after a short illness, and the Lincei had lost their leader. Castelli feared or knew about machinations of certain opponents and recommended having the work printed in Florence. Now Riccardi was startled and demanded to review again the manuscript corrected by Galileo. The latter, however, thought that this was unreasonable due to the state of the postal system. Then it was agreed that Galileo would submit only the introduction and conclusion in Rome, while the rest would be submitted to the inquisitor in Florence. Riccardi commended the affair to him in a long letter dated May 24, 1631. Since it is important for the course of the future trial, it is reprinted here verbatim.

> Mr. Galilei is thinking of publishing there a work of his, formerly entitled *On the Ebb and Flow of the Sea,* in which he discusses in a probable fashion the Copernican system and motion of the earth, and he attempts to facilitate the understanding of that great natural mystery by means of this supposition, corroborating it in turn because of this usefulness. He came to Rome to show us the work which I endorsed, with the understanding that certain adjustments would be made to it and shown back to us to receive the final approval for printing. As this cannot be done due to current restrictions on the roads and the risks for the manuscript, and since the author wants to finalize the business there, Your Very Reverend Paternity can avail yourself of your authority and dispatch or not dispatch the book without depending in any way on my review.

> However, I want to remind you that [the Holy Father] thinks that the title and subject should not focus on the ebb and flow but absolutely on the mathematical examination of the Copernican position on the earth's motion, with the aim of proving that, if we remove divine revelation and sacred doctrine, the appearances could be saved with this supposition; one would thus be answering all the contrary indications which may be put forth by experience and by Peripatetic philosophy, so that one would never be admitting the absolute truth of this opinion, but only its hypothetical truth without the benefit of Scripture. It must also be shown that this work is written only to show that we do not know all the arguments that can be advanced for this side, and that it was not for lack of knowledge that the decree was issued in Rome; this should be the gist of the book's beginning and ending, which I will send from here properly revised. With this provision the book will encounter no obstacle here in Rome, and Your Very Reverend Paternity will be able to please the author and serve the Most Serene Highness, who shows so much concern in this matter. I remind you that I am your servant and I beg you to honor me with your commands.[174]

With that the Theologian of the Pontifical Household adopted a verbal arrangement that Galileo himself had hit upon already in 1618 in a letter to Archduke Leopold of Austria. The only difference was

[174] Favaro, *Le Opere di G. Galilei*, 19:327; English text from *The Trial of Galileo: Essential Documents*, trans. Maurice A. Finocchiaro (Indianapolis: Hackett, 2014), chap. 4, §16 "Riccardi to Florentine Inquisitor (May 24, 1631), pp. 113–114.

that Galileo had written in that way ironically, while Riccardi took the words seriously!

On September 21, 1631, therefore, the inquisitor and the vicar general of Florence granted the imprimatur; the introduction and conclusion had been sent from Rome on July 19.[175] Riccardi, of course, must have been pressured repeatedly to do it, since he saw very clearly that he had got mixed up in a dubious affair, while Msgr. Ciampoli intimated to him that the pope himself wished the *Dialogo* to appear in print.[176] But now when Galileo added to the Florentine imprimatur the one from the Theologian of the Pontifical Household, Riccardi must have realized that his name had been misused. It is not surprising that he therefore had the first copies of the book that arrived in Rome in May 1632 confiscated at the customs office.

Nevertheless the book started to circulate among the public and made a sensation. From Fra Bonaventura Cavalieri, Fra Fulgenzio Micanzio, Fra Tommaso Campanella, and naturally from Castelli — all of them members of religious orders — Galileo received enthusiastic letters, which also praised above all his outstanding literary achievement.[177]

The trial

While Galileo was still complaining in Rome — for instance to Cardinal Barberini[178] — about the confiscation of his book, his opponents,

[175] See Riccardi's Accompanying Letter to the Inquisitor of Florence, July 19, 1631, in Pagano and Luciani, *I Documenti del processo di Galileo Galilei*, 113. The text of the foreword, the linguistic form of which was left to Galileo's discretion, is in ibid., 110–112.

[176] See D'Addio, "Considerazioni," 91; Eszer, "Niccolò Riccardi," 437–438.

[177] See Müller, *Der Galilei-Prozess*, 122–124.

[178] About Cardinal Francesco Barberini (1597–1679), the nephew of Urban VIII, see A. Merola, in *DBI* 6 (1964): 172–176.

whom he had so often and so offensively defied, were at work bringing about an ecclesiastical ban of the book. Friends and foes therefore formed two determined pressure groups, which opposed one another and sought to influence the course of events, each in its way.[179] Personal relations and motives, rivalries between religious orders, intra-bureaucracy tensions — all this would now work together to bring the Galileo affair onto the path which ultimately led to the recantation before the Inquisition.

Meanwhile the pope acted. In mid-August he ordered the appointment of a commission, with his cardinal-nephew presiding, that was to deal with the Galileo affair. No doubt the pope, for all the esteem that he still had for Galileo, was furious about the cooperation between Riccardi, Ciampoli, and Galileo, and when Francesco Niccolini, the Tuscan ambassador in Rome, asked the pope to give Galileo an opportunity to justify himself or to dispel misunderstandings, the answer he received was that Galileo had been informed very exactly by the pope himself about the difficulties that stood in the way of his statements. When Niccolini insisted, the pope said that the suppression of the *Dialogo* was the least that Galileo could now expect.

Nevertheless, he did not initiate the normal procedure by commending the affair to the Holy Office; moreover, he appointed his nephew as the president of the commission. This indicated that he still was trying to spare Galileo. Riccardi, too, wanted this and worked to achieve a prohibition of the *Dialogo* that was restricted in much the same way as the one that had been pronounced in 1616 against Copernicus, so that after corrections had been made the prohibition could lapse.

[179] See M. D'Addio, in *DBI* 6 (1964): 77, who on pages 78–91 also presents the pertinent maneuvers.

It is not known who the members of that commission were. On the other hand, we do know the result of their investigations, which is articulated in eight points:

1. Galileo put the Roman imprimatur at the head of his work without authorization and without sending the book to the censor who allegedly signed it.

2. The foreword printed in special type seems to be separated from the whole in such a way that it is utterly worthless for the purpose intended by the official censor. The final refutations are in lines assigned to a foolish man (Simplicius), in such a way that they can scarcely be found; moreover, they are received very coolly by the other participants in the discussion — often their effective sides are indicated only obscurely and with a certain reluctance.

3. Often the work no longer speaks about a mere hypothesis, in that either the movement of the earth and the immobility of the sun are simply asserted, or else the arguments for them are described as valid and necessary, while the contrary is considered impossible.

4. Galileo treats the question as one that has not yet been decided, as though we were only expecting, but not presupposing, a decision about them.

5. It is remarkable how roughly the opponents are treated — opponents who are often figures whose writings the Church makes particular use of.

6. Moreover, a certain equality that supposedly exists between human and divine knowledge with reference to geometric proofs is poorly explained.

7. It is maintained as a truth that followers of Ptolemy would probably become Copernicans, but not vice versa.

8. Furthermore the constant ebb and flow of the sea is inaccurately explained by the stationary position of the sun and the movement of the earth.[180]

The decisive point, however, was the third, which accused Galileo of having presented the heliocentric system as something conclusively proven.

It is also undeniable that Galileo treated those who thought differently in an arrogant, insulting way. For example, he spoke about Tycho Brahe's garrulousness, Kepler's childish observations, and Scheiner's fantasies. Point 5 is an objection to this way of dealing with his opponents.

Above all, however, another consideration weighed heavily in the scales: the disregard for the prohibition issued to Galileo in 1616 that was contained in the *registratura* dated February 26, 1616, which we have already discussed. If in fact the explanation is correct that at that time Bellarmine told Galileo that he did not have to take seriously this prohibition of the overzealous and hasty commissioner of the Holy Office, this could no longer be proved after Bellarmine's death.[181] In any event, the prohibition went into effect fully now.

[180] Müller, *Der Galilei-Prozess*, 131–132; Pagano and Luciani, *I Documenti del processo di Galileo Galilei*, 108.

[181] See Drake, "Galileo and the Church," 254–255.

With the approval of the pope, the commission now handed the affair over to the Holy Office, which in September 1632 started to investigate the *Dialogo*. The experts appointed were the theologian of the Holy Office, Agostino Oregio, the Jesuit Melchior Inchofer, and the Theatine Father Zaccaria Pasqualigo.

They came to the same conclusion the commission had, whereupon the Holy Office on September 23, 1632, made the decision to summon Galileo to Rome.[182] The party concerned received news of this on October 1.[183]

By now, though, Galileo was already seventy years old. His health was failing, and he balked at the wearisome journey.[184] Therefore, through Michelangelo Buonarroti, the nephew and namesake of the great artist, he sent to Cardinal Barberini the request to try his case in Florence.[185] Again the Tuscan ambassador, Niccolini, was the one who sized up the situation correctly and advised Galileo in late October to have no illusions about the outcome of the case. He should not try to justify his book, but rather should simply submit to the expected sentence of the Holy Office. He would not get around a trial and restrictions on his freedom.

When all attempts by Galileo to avoid traveling to Rome failed because the Holy Office and the pope himself would not bend, he started his journey on January 20, 1633.[186] On account of a plague that was raging in Florence, he had to be quarantined for twenty days

[182] See Pagano and Luciani, *I Documenti del processo di Galileo Galilei*, 113–114.

[183] See ibid., 115–116.

[184] See Inquisitor of Florence to Cardinal Barberini, November 20, 1632, in ibid., 117.

[185] See Galileo to Cardinal Barberini, Florence, October 12, 1632, in ibid., 118–119.

[186] See the report by the Inquisitor of Florence dated January 22, 1633, in ibid., 123.

at the border of the Papal States near Acquapendente; therefore he arrived in Rome on February 13 and took up lodgings in the Palazzo Medici, the residence of the Florentine ambassador. Cardinal Barberini advised him not to receive visitors, since that could complicate his situation. On the other hand, one of the consultors of the Holy Office, Msgr. Ludovico Serristori, visited him twice, in order to hear his arguments. He too recommended that Galileo show a willingness to submit. The fact that he was not taken into custody but rather was allowed to live in the Palazzo Medici was perceived by his contemporaries as quite unusual. For this reason Galileo himself also had the feeling that the storm had meanwhile been calmed. He described the way in which Msgr. Serristori had met with him as very amiable and obliging, and he took it as a sign that the authorities intended to let kindness and leniency prevail.

In Rome opinions were divided, and this became increasingly evident. In particular, Castelli was eager to win the competent parties over to Galileo's side.[187] The influential Lucas Holstenius, who after his conversion in the year 1625 lived in Rome and was appointed Apostolic Prothonotary by Urban VIII — later he became prefect of the Vatican Library — expressed his opinions in complete agreement with Galileo. Even the Commissioner of the Holy Office, Father Vincenzo Maculano, took his side, which was true of Cardinals Capponi and Scaglia, also. Meanwhile Niccolini did what he could to convince the pope but was unable to dissuade him from his decision to summon Galileo before the Holy Office. That, he thought, was the least that ought to happen. Nevertheless, the pope promised that the best and most comfortable rooms would be assigned to Galileo. The astronomer had to move into them on April 12; they were part of the residence of a high-ranking official of the Holy Office, who had

[187] See D'Addio, in *DBI* 6 (1964): 92–95.

vacated them for him. Fr. Maculano welcomed him there in a markedly friendly way and also said to him that he was allowed to move freely in the house and garden. He was permitted to bring his servant with him; meals were brought to him from the kitchen of the Palazzo Medici. He could also engage in free and unrestricted written correspondence.

On that day Galileo was also interrogated for the first time by Fr. Maculano.[188] It was certainly difficult for Maculano to perform this task, because in the preceding year he had expressed to Castelli his conviction that the heliocentric system was altogether acceptable. Now, though, it was not a matter of his personal conviction but rather of his office.

The first questions revolved around the proceedings in 1616. Galileo answered them as follows: "One morning His Eminence Cardinal Bellarmine called for me and said something definite to me that I would prefer to tell His Holiness first personally, before I said it to others. Finally he told me that it is not possible to hold or to defend the opinion of Copernicus, since it contradicts Sacred Scripture."[189] When questioned further, he gave this answer: "It may be that some command was given to me that I must neither hold nor defend the aforementioned opinion, but I no longer know for sure, since this was several years ago now."[190] Neither could he remember where he had been forbidden to teach Copernicus's opinion "in any way whatsoever"—that was the wording of the *registratura* dated February 26, 1616.

Then the interrogation turned to the way in which the imprimatur for the *Dialogo* had come about. Galileo let himself get

[188] The protocol of the interrogation, signed in his own handwriting, is in Pagano and Luciani, *I Documenti del processo di Galileo Galilei*, 124–130.

[189] Ibid., 127.

[190] Ibid., 127–128.

carried away on this topic and stated that he had told Fr. Riccardi nothing about the prohibition in 1616, because in this book he neither advocated nor defended the heliocentric system. It was plain to everyone involved that this was a falsehood, but it agreed with the foreword to the *Dialogo* that had been arranged with Fr. Riccardi; of course, the contradiction between it and the contents of the book had not escaped the experts. Fr. Inchofer in particular had pointed it out.

Now this statement made everything more difficult, and Galileo's situation deteriorated, especially since Galileo had testified under oath.[191] If the Holy Office or even the pope had intended to destroy Galileo, this alone would have given them the ways and means of doing so.

Instead, Fr. Maculano requested permission to speak with him privately, so as to investigate other possible ways to proceed; in particular he tried to move him to admit his "error," and he succeeded. After that, Maculano told him that the main charge would now be that, contrary to the 1616 prohibition, he had advocated Copernicus's teaching in the *Dialogo*. As soon as he admitted that, he would be asked about his real opinion and an opportunity would be given to him to defend it, and then he could be dismissed to go home — that is, to the Palazzo Medici. This private conversation on April 27[192] was the preparation for a further judicial interrogation on April 30.[193]

Galileo had meanwhile sent for a copy of his work and said that after three years he had read it again, this time very carefully, and

191 See D'Addio, in *DBI* 6 (1964): 98–102.

192 About this see Maculano's report to Cardinal Barberini, Rome, April 28, 1633, in Favaro, *Le Opere di G. Galilei*, 15:106–107).

193 Protocol with Galileo's own signature is in Pagano and Luciani, *I Documenti del processo di Galileo Galilei*, 130–132. On the same day he was released to his new lodgings, the Villa Medici; see ibid., 133.

that it seemed quite strange to him, like the book of a different author. Now he himself realized that in many passages the book could give the impression that he shared the opinion of Copernicus. The fact that he formulated the reasons in favor of it more impressively than the arguments against it could be blamed on his vanity as a scholar.[194] Moreover, he volunteered to write a sequel to his work, which would clarify all the open questions. Thereupon Galileo returned to his quarters.

Meanwhile Galileo was preparing a written defense, which was based chiefly on the fact that he forgot Bellarmine's verbal prohibition against teaching Copernicus in any way and adhered to the wording of the decree from the Congregation of the Index, which did allow a hypothetical treatment of Copernicus. Consequently they could probably not accuse him of disobedience.[195]

In this way Galileo did comply with the effort by his judges to guide the unpleasant case to a good outcome. Yet it could not remain hidden from them, either, how nimbly Galileo turned against what he had advocated. After all, they had his sworn statement from April 12 in the files.

Therefore, the cardinals could not simply acquiesce in this contradiction, unless they wanted to make themselves Galileo's accomplices. For this reason various expert opinions were demanded

[194] On this subject, see the interesting comment: "In the final analysis the problem the *Dialogue* created for Galileo was that his rhetoric worked too well. His arguments in the fictional format were effective — his drama succeeded in showing that the arguments in favour of Copernicus were the most compelling and that those of the peripatetics were ridiculous. What failed was the ethos he projected in the framing of his drama, and only Galileo really knows, whether it was ultimately misjudged" (J. D. Moss, "Galileo's Rhetorical Strategies in Defense of Copernicanism," in Galluzzi, *Novità celesti*, 103.

[195] See statement dated May 10, 1633, in Pagano and Luciani, *I Documenti del processo di Galileo Galilei*, 135–137.

on the question of whether or not Galileo taught the movement of the earth in the *Dialogo*. The authors of three opinions — the aforementioned Consultor of the Holy Office (and later cardinal) Agostino Oregio, the astronomer Melchior Inchofer, S.J., and the theologian Zaccaria Pasqualigo (the latter two substantiated their judgments in great detail) — unanimously agreed that in the *Dialogo* Galileo taught and defended the Copernican system.[196]

Therefore a further interrogation was scheduled for June 21, whereby Galileo was to be questioned about his true opinion, even under threat of torture. He answered that before 1616 he had wavered between the two cosmologies. "After that Decree in 1616," he continued, however, "all doubt within me vanished, and I held, as I still do hold also, the teaching of Ptolemy, *i.e.* that the earth stands still and the sun moves, to be altogether correct and indubitable."[197]

Thereupon they called his attention to the obvious contradiction between this statement and the overall intent of his *Dialogo* and admonished him — and this was a procedural formality provided for by the law — to tell the truth, while referring to the possibility of torture. Galileo, however, stood by his statement, and after that they had him sign the protocol.[198] With that the strategy that had been pursued, especially by Maculano, to wrap up the trial and thus to avoid the perplexing situation that would necessarily arise if a book that had appeared with an imprimatur was subsequently forbidden, had failed definitively.[199]

The great, gloomy, oft-invoked concluding scene played out on the following day, June 22, 1633, in an assembly room of the

[196] About these texts — the one by Oregio is dated April 17, 1633 — see ibid., 139–153.

[197] Ibid., 155.

[198] Protocol signed in his own hand is in ibid., 154–155.

[199] See D'Addio, in *DBI* 6 (1964): 106–107.

Monastery of Santa Maria sopra Minerva. There Galileo was informed of the judgment, the most important passages of which read:

> We say, pronounce, sentence, and declare that you, the above-mentioned Galileo, because of the things deduced in the trial and confessed by you as above, have rendered yourself according to this Holy Office vehemently suspected of heresy, namely of having held and believed a doctrine which is false and contrary to the divine and Holy Scripture: that the sun is the center of the earth's orbit and does not move from east to west, and the earth moves and is not the center of the world [i.e., universe], and that one may hold and defend as probable an opinion after it has been declared and defined contrary to Holy Scripture.
>
> Consequently you have incurred all the censures and penalties imposed and promulgated by the sacred canons and all particular and general laws against such delinquents. We are willing to absolve you from them provided that first, with a sincere heart and unfeigned faith, in front of us you abjure, curse, and detest the above-mentioned errors and heresies, and every other error and heresy contrary to the Catholic and Apostolic Church, in the manner and form we will prescribe to you.
>
> And, in order that this your grave and pernicious error and transgression may not remain altogether unpunished and that you may be more cautious in the future and an example to others that they may abstain from

> similar delinquencies, we ordain that the book of the *Dialogues of Galileo Galilei* be prohibited by public edict.
>
> We sentence you to imprisonment in the Holy Office at our discretion, and as a salutary penance we oblige you for three years to recite the seven penitential psalms once a week. Moreover we reserve the right to reduce, to change, or to rescind entirely or partially the aforesaid punishment and penance as we see fit.[200]

After that Galileo had to read aloud the formula of abjuration that was presented to him, the essential sections of which are reprinted here:

> I, Galileo, son of the late Vincenzo Galilei from Florence, 70 years old, having been put on trial personally and now on my knees before Your Eminences the Cardinals and Most Reverence General Inquisitors against the heretical corruption for all Christendom, holding in front of me the most holy Gospels, which I touch with my hands, do swear that I have always believed, believe now, and in the future with God's help intend to believe all that the Holy Catholic and Apostolic Church considers, preaches, and teaches to be true. I, however, was condemned by the Holy Office as being strongly suspect of heresy, because I wrote and published in print a book which discusses the teaching that has been condemned as false, that the sun is motionless at the center of the universe, while the earth moves outside of the

[200] Favaro, *Le Opere di G. Galilei*, 19:405–406; translated from the German in Müller, *Der Galilei-Prozess*, 152–153.

center of the universe, and I supported this teaching very effectively with arguments, without indicating the solution to them. I did this despite the precept that had been communicated to me officially, to give up that false teaching entirely, and neither to hold it as true nor to defend it or teach it in any way whatsoever in speaking or in writing. . . .

Now, since I wish to take away, both from Your Eminences and from every Christian believer, this strong suspicion that is rightly harbored against me, therefore I abjure, deplore, and despise the aforesaid errors and heresies, along with any other error, heresy, and sect whatsoever that is contrary to the Holy Church; I also swear in the future never again to try to say or affirm anything similar, either in writing or in speech, through which such suspicion about me could arise; but if I become acquainted with any heretic or anyone suspected of heresy, I will report him to the Holy Office, to the Inquisitor, or to the Bishop of the locality in which I find myself. . . .

I also swear and promise that I will perform completely and comply with the penitential works imposed on me, and should I, which God forbid, act contrary to any one of my sworn promises, then I submit to all the penances and punishments that are appointed and published by the holy canons and other general or particular constitutions against similar offenses.

> So help me God and these, His holy Gospels, which I touch with my hands....
>
> I, the aforesaid Galileo Galileo, have abjured, sworn, promised, and obliged myself, as was said above, and, as a testimony to the truth, have signed in my own hand the present document and have read it aloud word for word.
>
> In Rome, the convent of Santa Maria sopra Minerva, today, the 22nd day of June, 1633 A.D. I, Galileo Galilei have abjured as was said.[201]

The whole proceeding was not uniformly judged even within the Inquisition, as is demonstrated by the fact that three of the ten cardinals of the Holy Office did not sign the judgment, among them the pope's nephew Francesco Barberini. According to an order given by Urban VIII dated June 30, 1633, the aforesaid texts were to be made known to all members of the Holy Office as well as to all professors of philosophy and mathematics in an assembly that was to be convened specially, and also — by mail — to all nuncios, particularly those of Padua and Bologna, and also to the inquisitors.[202]

Here too, now, arises the oft-repeated question about Galileo's behavior in this situation that was so unfortunate for him.

Bertolt Brecht's judgment is well known. He has the character of Andrea scream in Galileo's face when he returns: "You drunkard! You

[201] Favaro, *Le Opere di G. Galilei*, 19:406–407; translated from the German in Müller, *Der Galilei-Prozess*, 155–156.

[202] See Pagano and Luciani, *I Documenti del processo di Galileo Galilei*, 156.

scavenger! Have you saved your beloved skin now?"[203] Another author speaks about Galileo's "altogether unfortunate declaration that shows a lack of character."[204]

In contrast, another judgment shows much more understanding for the religious dimension of the trial, which no doubt was clearer to Galileo than to contemporary man:

> It has been emphasized more than once that Galileo was interested in religion. The various formulaic phrases of the judgment and also of the abjuration formula point to the ecclesiastical realm. The judgment and abjuration took place on ecclesiastical premises. Even a Christian today who is loyal to the Church must therefore view the event with different eyes than a freethinker around 1900 might; how much more, then, a seventeenth-century Catholic! Galileo repeatedly emphasizes that he is there in order to show obedience. If we consider that the point of departure of the trial for him is the fact that he wants to keep the Catholic Church from making a far-reaching wrong decision, then it would be illogical if finally he no longer cares about his Church. Galileo does not want to be a heretic; for him this is a horrible accusation, probably less on account of the consequences that may affect him on earth than because he is concerned about eternal happiness. For many people around the year 1900 it was probably quite impossible to picture themselves in the

[203] Bertolt Brecht, *Werke, Stücke* vol. 5, *Leben des Galilei*, 1955–1956 version (Berlin-Weimar-Frankfurt am Main, 1988), 274.

[204] Hemleben, *Galileo Galilei in Selbstzeugnissen*, 123.

> time 300 years earlier. But if we imagine Galileo torn by the dilemma of choosing between eternal damnation and his own rational thought, then the decision can never be the one that a man around 1900 makes. What seems later to be repulsive in his behavior, indeed lacking in character, is suddenly understandable as the interior tragedy of a man who very deeply despairs at the gulf that is opening up between religious faith and rational thought. Thus even the unreliable statement that he, Galileo, was not convinced of the truth of the Copernican teaching when he wrote his *Dialogo,* becomes understandable as the tormenting question he asked himself: Have I really become so unfaithful to the Church that I have deliberately promoted heresy? No, that is completely impossible! I was still a loyal Catholic, and therefore I complied with the instructions that were dictated to me! Therefore I would like to see Galileo as a faithful churchgoer who went to confession regularly, trapped in a hopeless dilemma that honors him and does not demean him along the lines of falling away from a conviction. He does not intend to sin and yet suspects that he is sinning. Consequently, though, any consistent behavior for him in the Inquisition trial now becomes downright impossible.[205]

This interpretation, which is probably overdramatic and plainly should be understood in terms of the author's Protestant experience, nevertheless does not clarify everything. The fact that Galileo testified under

[205] Freiesleben, *Der Prozess gegen Galilei,* 23–24.

oath that he never considered the heliocentric system true, is no doubt grave. Maybe the pressing situation in which he found himself prevented him from judging this conduct clearly. But it is very dubious, indeed quite unlikely, that he acted against his better knowledge and conscience in the abjuration itself, although he certainly did not give up his opinion about the earth's movement and the sun as the center of the universe. The latest analyses of Galileo's *Dialogo,* together with its early Latin manuscripts, have shown unambiguously that in no passage of his *Dialogo* did Galileo claim to have furnished conclusive proof for the Copernican theory.[206] Moreover he was certainly aware that such a demonstration was problematic in terms of the theory of science. Thus he also clearly realized that his conviction lacked final validation.[207] An abjuration would have to be viewed then as a consequence of this state of affairs.

But even if Galileo had been subjectively quite sure of his point, he could have signed the recantation in good conscience. Indeed, Galileo was well instructed theologically, and also too well advised by his numerous priest friends to be incapable of judging soberly and accurately the import of the event on June 22. He was very well aware of the difference between the infallible judgment of a pope or a council on a matter of faith and a judgment of the Holy Office. Whereas a decision by a pope or a council about an article of faith in fact concerned the Christian's innermost assent of faith, a Roman dicastery like the Holy Office, even if the pope presided, could not require this innermost assent of faith, though it could demand the obedience that a Catholic owes to the so-called authentic (i.e., not absolutely infallible) Magisterium. Galileo

[206] See William A. Wallace, "Galileo's Concept of Science: Recent Manuscript Evidence," in *The Galileo Affair,* 34.

[207] See William A. Wallace, "Galileo and Aristotle in the 'Dialogo,' " *Angelicum* 60 (1983): 311–332. Galileo's brilliant rhetoric was supposed to "fill in" the gaps in the proof; this is demonstrated by J. Dietz Moss, "The Rhetoric of Proof in Galileo's Writings on the Copernican System," in *Galileo Affair,* 41–65.

himself often spoke and wrote about his willingness to obey in this way, and it is not possible to dismiss statements of this sort as hypocrisy.[208]

Such an interpretation may be accurate. Another argument in favor of it is Descartes's behavior after he composed his book *Le monde* (The World), which was likewise based on a Copernican foundation; he found himself in a situation similar to Galileo's. A few months after his condemnation, in April 1634, he wrote to his friend Mersenne, who was trying to force him to publish *Le monde*:

> Although I thought that they [i.e., Galileo's ideas concerning the earth's movement, etc.] were based on quite certain and evident proofs, nevertheless not for anything in the world would I want to cling to this view against the authority of the Church. I know, of course, you could say that a matter that has been decided by the Roman Inquisitors does not yet thereby become an obligatory article of faith.... But I am not so fond of my own thoughts that I would want to cite such restrictions just to hold on to them.... But since I have never read that this censure was authorized by the Pope or by a council, but only by a single Congregation of Cardinal-Inquisitors, I do not give up all hope that in this matter something similar may happen as with the *Antipodes*, which were also once more or less forbidden. Then over the course of time my book *Le monde* too could see the light of day. I will have to apply all my intellectual faculties to bring about this situation.[209]

[208] See Soccorsi, "Il Processo di Galileo," 888–895.

[209] Descartes to Mersenne, Amsterdam, April 1634, in *Oeuvres de Descartes*, vol. 1, ed. Charles Adam and Paul Tannery (Paris, 1897), 285, 288). See also D'Addio, in *DBI* 6 (1964): 113–115.

Therefore, we should assume that Galileo understood his abjuration as an act of loyalty to the Church, which he could perform in good conscience, albeit with a heavy heart, without having to give up his personally held scientific conviction, however well substantiated he might consider it to be. Of course, the fact that from then on he could no longer advocate it in his words and writings was enough to bear.

If we have tried to put ourselves in Galileo's place, then it is appropriate to make the same attempt with regard to his judges. Generally they meet with a clear condemnation by later commentators. For example, there is nothing ambiguous about this statement: "Galileo lies, he must lie — anything else would have cost him his life. No one has the right to judge him, unless he himself has been in a similar situation, for example in the hands of a *Gestapo*."[210] A little further on the same author opines that Galileo's opponents "made use of the institution in order to carry out their power plays with theological trimmings."[211] Their goal was "to suppress thought."[212] Such judgments are widespread; it is enough to refer to our introductory chapter.

Until now, though, hardly anyone has taken the trouble to inquire seriously and without bias into the self-understanding and the presuppositions of Galileo's judges. They have fallen victim to the same general verdict as leveled against the Inquisition to which they belonged — a judgment that has long since been considered self-evident.

More recent investigations make it possible, indeed necessary, to judge the Inquisition objectively and accurately — even its record

[210] Hemleben, *Galileo Galilei in Selbstzeugnissen*, 123; "Ein Schandurteil...," *ibid.*, 9.

[211] Ibid., 129.

[212] John Carter and Percy H. Muir, *Printing and the Mind of Man* (New York: Holt, Rinehart & Winston, 1967).

during the medieval era.[213] According to these studies it is not possible to imagine the Inquisition as a group of power-hungry, sadistic, blindly raving, gloomy fanatics, who were concerned about cudgeling free thought into submission since they viewed it as endangering the Church's position of power. On the contrary, here again is another, obviously non-Catholic opinion: "It was by no means true that power-hungry clerics as a matter of principle had closed their minds to new findings here. There is indeed a lot of ugliness to record over the course of the trial; but to deny the lofty sense of responsibility of the great majority of the clergy involved with the matter would be simply unhistorical."[214]

However, what distinguishes the inquisitors fundamentally from us and what makes it almost impossible for people today to understand them is the hierarchy of values in which their institution was rooted and by which they oriented their decision. For them, the highest position was held by the divinely revealed truth of the faith which had been entrusted to the Church to preserve and proclaim. This was an absolute value and an absolute standard. Consequently there were narrow limits to the autonomy of earthly, human values. Now if some human knowledge is recognized as contradicting divine revelation, then there are only two possibilities: either the contradiction is only apparent, or else man's knowledge is based on error, because the truth cannot contradict the truth, whether it comes from nature or from revelation. What necessarily was left behind in this

[213] See the numerous studies on the topic "Inquisition" by Alexander Patschovsky: e.g., *Die Anfänge einer ständigen Inquisition in Böhmen: Ein Prager Inquisitoren-Handbuch aus der ersten Hälfte des 14. Jahrhunderts,* Beiträge zur Geschichte und Quellenkunde des Mittelalters 3 (Berlin-New York: De Gruyter, 1975); "Strassburger Beginenverfolgungen im 14. Jahrhundert," *Deutsches Archiv* 30 (1974): 56–198; and "Zur Ketzerverfolgung Konrads von Marburg," *Deutsches Archiv* 37 (1981): 641–693.

[214] Freiesleben, *Der Prozess gegen Galilei,* 3–4.

rigorous ethos of truth was the freedom of the human being who is conducting an investigation and also his right to search by trial and error. This ethos, of course, led to restrictions only with regard to the publication of such thoughts. But even that is for us today an intolerable idea, since for us, in direct opposition to the medieval and Baroque view of this problem, human, individual freedom — and indeed limitless freedom — has become the highest value and standard. Modern man is reluctant, of course, to admit that this system, too, necessarily has its disadvantages.

At this point we should examine the judgment of Galileo's judges by Jacques Maritain, who accuses them of serious abuse of authority because they used extreme moral coercion. The reprehensibility of their proceedings rests on the fact that everyone, but especially the judges themselves, knew that their doctrinal decision could make no claim to infallibility, neither with regard to its object nor with regard to the limited competence of those involved. Now their guilt consisted in the fact that they forced Galileo to abjure heliocentrism, which they had so irrevocably declared to be contrary to Scripture, as a heretical teaching. More sensibly, they should have issued a prohibition of Copernican propaganda. On account of this state of affairs — as Maritain thinks he can see it — he accuses Galileo's judges of arrogance and authoritarianism. With this indictment, however, his purpose is not to belittle the individual bureaucrats of the Holy Office, whose personal humility he does not mean to deny, but rather to describe the psychology of the dicastery to which they belonged.[215]

[215] See Jacques Maritain, *De l'Église du Christ: La personne de l'Église et son personnel* (Paris-Bruges: Desclée de Brouwer, 1970), 357–360. A good example of a philosopher's unhistorical way of looking at things! Even the "psychology" of the Holy Office that he alleges nevertheless depicts an element of the historical circumstances of the trial.

However, a look at the wording of the authentic texts would have kept the philosopher Maritain from making this erroneous judgment: no one ever demanded that Galileo abjure the heliocentric system as a heresy!

Whenever we are willing to make a historical judgment objectively, then, let us concede that Galileo's judges used the standards of their own convictions and of their own time. Then, of course, we will not be able to overlook the fact that they strove for justice, as far as it concerns both the procedures and also the judgment itself. In any case, we can clearly observe an antagonism among the personnel within the Holy Office, whereby some—think for instance of Fr. Maculano and several cardinals—were sympathetic to Galileo, while naturally the adherents of the Peripatetic school also had their representatives. Possibly in their ranks could be found an aversion to Galileo, who had just insulted and challenged them in a way that could scarcely be tolerated any longer.

Moreover Galileo had made it difficult, even for the judges who were well disposed toward him, to help him, when he confronted them with such obvious untruths and excuses (as anyone who had read the *Dialogo* would discover immediately). At the final interrogation on June 21, the court refrained from reproaching him about the long list of passages from the *Dialogo* which would have demonstrated conclusively the falsehood of his statement; this can only be understood, therefore, as the judges' attempt, made almost against their better knowledge, not to carry the matter too far.

An opinion already cited spoke about the "ugliness" that had come to light over the course of the proceedings. What is that supposed to mean? On the one hand the author got the impression that the judgment had already been fixed before the hearing. But then he also speaks about the pope's direct intervention into the proceedings and about arbitrariness.[216]

This last accusation originates with the idea of the division of powers, as it is put into action in a democratic society and is expressed in the independence of a judge and of a court. Such notions, and therefore also such a standard, are of course foreign to the seventeenth century and to the Church's canon law in general. Every ecclesiastical judge is a mouthpiece of the one invested with pastoral authority who appointed him — that is, in Galileo's case the supreme judge was the pope. His immediate intervention into the proceedings, therefore, can only be offensive to a modern observer. One could, though, speak in a certain way about arbitrariness and about a preconceived judgment — of course this happened instead in Galileo's favor. Indeed, if they had dealt with him strictly according to the law, then at the very least they would have had to indict him for his manifestly false statements. Therefore they had probably decided to spare Galileo personally, as far as possible, while at the same time making an example of him, if you will, that would inculcate respect for the 1616 decree of the Congregation of the Index.

The way in which the sentence against Galileo was carried out could support the interpretation attempted above. As early as June 23, by order of the pope, the Palazzo Medici was assigned to him as his prison. One week later Urban VIII decided that Galileo should go to Siena and live there in the house of Archbishop Ascanio Piccolomini.[217] In Florence the plague was raging at that

[216] See Freiesleben, *Der Prozess gegen Galilei*, 16 or 22.

[217] See the note from the files dated June 30, 1633, in Pagano and Luciani, *I Documenti del processo di Galileo Galilei*, 156; or July 2, 1633, in ibid., 156–157. With it the pope replied to a request by Galileo dated June 30, 1633, who referred to the lack of space in the Villa Medici, where many visitors were expected (ibid., 157).

time. Piccolomini was a student and close friend of Galileo. For him it was a pious duty to make the stay as encouraging as possible for the old man, who was shaken to the core by the events in Rome. Galileo himself reports that Piccolomini treated him like a father and that he constantly had distinguished visitors from Siena around him. The evidently pleasant life that Galileo was therefore able to live there, though, angered some citizens of Siena who were ill disposed toward him, and they filed a complaint with the Inquisition. But neither that institution nor Urban VIII reacted to it.[218]

In conversation with Ambassador Niccolini, the pope said that he intended to go step-by-step with Galileo's rehabilitation, and so it happened. At Galileo's request he was permitted on December 1 "to return to his country seat; nevertheless he should live there quietly and not host any gatherings or give conferences until further notice from His Holiness."[219] It is often maintained that he was under arrest, but it certainly cannot be called that.

In Arcetri he was welcomed with tumultuous joy by his daughter Maria Celeste, in which the prioress and the sisters of her convent also shared cordially. It is indicative of the relation between father and daughter that Maria Celeste immediately took over the penance imposed on her father, to pray once weekly the seven penitential psalms.[220] In thanksgiving for their amicable assistance in the past eleven difficult months, Galileo sent to the wife of Ambassador Niccolini a valuable cross, and to the archbishop of Siena a telescope.[221]

[218] See Paschini, *Vita e opere*, 559; Drake, "Galileo and the Church," 354–356.

[219] Müller, *Der Galilei-Prozess*, 158.

[220] See ibid., 167.

[221] See Paschini, *Vita e opere*, 561.

The harvest of a life of research

It is a sign of great bodily and intellectual vitality that only after the bitter experiences in Rome and on the threshold of his seventieth year Galileo began the creative period in which he established his real importance in the field of physics.[222]

First it took a while before Galileo, thanks to his close household fellowship with Archbishop Piccolomini in Siena, regained his equilibrium of mind and soul. Although at the beginning he had wandered about the episcopal palace at night, restless and crying out, his host soon succeeded in guiding his teacher's thoughts back to science. Contrary to a common assumption, Galileo did not occupy himself now with writing down the "Third Day" of the *Discorsi*: rather, the structure of matter and the solidity of materials captured his attention and were now the object of his scientific correspondence — preliminary studies for the "Second Day" of the *Discorsi*. At the same time he set out to formulate parts of the "First Day," whereby he made use of the dialogue genre, of which he had such a masterful command. This text dealt for example with the behavior of liquids, leading to reflections on the structure of matter, about which Galileo was debating with Alessandro Marsili, a professor of philosophy in Siena. This emerges from his correspondence during that period and from a relevant communication of Piccolomini to Galileo's later biographer Vincenzo Viviani.

After Ambassador Niccolini has already made an effort on November 13, 1633, to obtain the pope's permission for Galileo to return to Florence, on December 1, Urban VIII allowed the astronomer to go to Arcetri near Florence, where Galileo owned a villa, *Il*

[222] See in general Alistair Cameron Crombie, *The History of Science from Augustine to Galileo* (New York: Dover, 1996); Drake, *Galilei at work*; Freiesleben, *Galilei als Forscher*.

Gioiello ("the jewel"). In the Convent of San Matteo, located near Acetri, his favorite daughter Virginia was living as Sister Maria Celeste. Her unexpected, early death in April 1634 deeply wounded the father. Even his physical condition suffered much as a result; in the following period also health problems and wellness alternated again and again. Galileo's eyesight had become weaker even before that, and now in 1637 he went blind in both eyes. For three months he had no desire whatsoever to do scientific work.

Yet Galileo was not alone and abandoned. He was able to gather around himself students, among them several young clerics from the Piarist order, which was dedicated to educating youth. Above all, though, Viviani, who then also became his master's biographer, stayed with him for two years. Soon Galileo was also receiving frequent visits from his friend Castelli who, considering his special relations, had asked the pope for permission to make these visits and had received it.

Apart from these trusted friends and students, persons of great renown also made their way to Arcetri. Besides the members of the Grand Duke's family, both John Milton and Thomas Hobbes were guests there, and probably Descartes, too. During the final period, Evangelista Torricelli too was staying in Arcetri.

Among these acquaintances, Galileo continued his projects with great inner commitment. Upon his return from Siena he had found in Arcetri a letter from Professor Matthias Bernegger of Strasbourg, who had prepared Latin versions of several other works by Galileo, offering to translate the *Dialogo* also. In that same year Prince Mattias de' Medici, then governor of Siena, brought the Italian original to Strasbourg, where Bernegger took care of the translation.[223] It was

[223] On this topic and on Vienna's role in this connection, see Zdenko Šolle, *Neue Gesichtspunkte zum Galilei-Prozess* (Vienna: Verlag der Österreichischen Akademie der Wissenschaftern, 1980), 51–53.

published by Elzevier in Leiden, with which Galileo remained in constant contact via several middlemen. Thus the same house printed his hitherto unpublished letter to Madame Christina. He also negotiated an edition of his complete works, whereby he made it clear that the *Dialogo* would be omitted.

Finally, Galileo was conducting negotiations with the Dutch government concerning the determination of geographical longitude at sea, as he had done two decades before. Nevertheless, this time, again, nothing was accomplished.

It is astonishing and admirable, however, that under the circumstances just described Galileo was able to crown his life's work in physics with the completion and publication of the *Discorsi.* The work was published in 1638, again by Elzevier, under the title *Discorsi e dimostrazioni matematiche intorno a due nuove scienze attenenti alla mecanica e i movimenti locali.* (Discourses and mathematical proofs concerning two new sciences pertaining to mechanics and local movements). As with the 1632 *Dialogo,* it was composed in the form of a learned debate, whose participants have the same names as the ones in the *Dialogo*. Their standpoints, too, remain as before. Constructed in four sections, each covering one day of the discussion, the work reaches the climax of its scientific message in its third and fourth "days." Here, uniform motion (falling), naturally accelerated motion (projectiles), the movement of a pendulum, solidity, and impetus are discussed and presented. The work was already recognized by mathematicians and physicists in the second half of the seventeenth century for its importance as an introduction to a new era in the teaching of mechanics. Building on the foundations laid here, Newton was able to formulate the concept of gravitation, to prove its validity for the heavenly bodies, and thus to produce a proof for Copernicus. This is probably Galileo's own true achievement for the continued development of astronomy and cosmology. With this work Galileo established, not his renown, but rather his historical importance for the progress of knowledge in the field of physics.

It is indicative that the Inquisition and the pope remained silent about this astonishing scientific activity of a "prisoner of the Inquisition" and let him alone. The *Discorsi* appeared in Leiden, naturally without the Church's imprimatur; when the first fifty copies arrived in Rome, they were immediately sold out, without any restrictions on their sale. Even the visitors from Protestant England met with no objections at all on the part of the Inquisition. These facts necessarily put into perspective Galileo's complaints about obstacles to his work.

Concerning Galileo's condition during this time we have a report of the Florentine inquisitor Muzzarelli to Cardinal Francesco Barberini about a visit in Arcetri on February 13, 1638.

> In order to comply as thoroughly as possible with the command of His Holiness, I unexpectedly went to Galileo's villa in Arcetri in the company of a physician from out of town with whom I am acquainted, in order to find out the state of his health.
>
> My purpose in doing this was not so much to be able to report on his ailments as to gain insight into the studies to which he is dedicating himself, and into the society that he keeps, so as to see to what extent he would be capable, if he were to return to Florence, of spreading his condemned teaching about the earth's movements in gatherings and speeches.
>
> I found him utterly devoid of sight and completely blind. Although he hopes for a cure, since it has been less than six months since he has suffered from

> cataracts, nevertheless the physician considers the ailment almost incurable, considering that the patient has just turned seventy-four. Moreover he has a serious hernia, which causes him constant pains, and suffers from insomnia, on account of which, as he himself and his fellow lodgers report, he can sleep less than one hour out of twenty-four. Moreover he is doing so poorly that he resembles a dead man more than one who is alive. The villa is far from the city in an inconvenient location, which is why he can only seldom call the doctor, which involves difficulties and major expenses. His studies were interrupted by his going blind, although now and then he has someone read something to him, and his company is not sought, because given the state of his health he usually can only complain about his misfortune and speak about his ailments, if someone does occasionally visit him. Considering this, I think that if Your Holiness in your infinite goodness were to allow him to stay in Florence, it would not be possible for him to host gatherings, and if he did, he is in such a remorseful disposition that I think a stern admonition would be enough to keep him in check. So much can I present to Your Eminence.[224]

Considering that as late as 1640–1641 Galileo engaged in a scientific controversy with his former student Fortunio Liceti about the light of the moon and in doing so deployed all of his skill at literary polemics, and furthermore that he gave instructions at his house to the intellectually not-so-gifted child of a certain Cesare Monti, the lines

[224] Favaro, *Le Opere di G. Galilei*, 17:290.

written by the inquisitor seem exaggerated, especially since Muzzarelli with this letter intended to obtain the pope's permission for him to move to Florence. The details about the state of Galileo's health, however, might be reported accurately.

In the autumn of 1641 Galileo became seriously ill; he felt the end of his life approaching. At his request, Castelli's most gifted student, Evangelista Torricelli, now came to Arcetri shortly before Galileo's death, and so did Castelli himself as he was traveling through from Northern Italy to Rome. In early January 1642 the state of his health deteriorated perceptibly. Encircled by his students and friends, among whom there were two priests, he finished his earthly course on January 8.

Both the productive activity of his old age, burdened as it was with sickness and blindness, and also the pious end of his life show that Friedrich Dessauer's tragic-sounding epithet "a failure" is groundless. Galileo was not a failure either as a researcher or as a Catholic when he was buried in Santa Croce in Florence.

Chapter 3
Causes and Connections

Persons and powers

After this rapid presentation of the events, which naturally is incomplete in its details, the urgent question arises about the factors that led to Galileo's disciplinary punishment by the Church. In the attempt to ascertain the causes and the background of a historical incident, we should carefully avoid the danger of explaining it with a single cause. When it is a question of human action, one should always start by assuming multiple factors at work causally, the effects of which are often so entwined and snarled that they can be untangled only partially, if at all. It is also difficult to decide which of them ultimately was decisive, so that the events played out this way and did not take some other course. With this proviso, then, let us get to work.

If we can believe him personally, it was above all else the enmity of the monks (in other words, the Dominicans and the Jesuits) that led to Galileo's fall. One of them, the Jesuit Grienberger, even seems to confirm this when he opines that Galileo could have lived in peace and enjoyed his renown if he had managed to keep the friendship of the Jesuits. This precise point has therefore been the object of various investigations. Now three Jesuits, of all people, defended themselves against this indictment of their order. However, neither polemical nor apologetic intentions count; arguments are the decisive thing. They are Fathers Hartmann Grisar and Adolf Müller, who wrote around 1900, and Filippo Soccorsi, whose study "Il processo di Galileo" is contained in volume 3 of the *Miscellanea Galileiana* of 1964. It contains also an excursus about Galileo's opponents.[225] They all declare that neither Fathers Grassi nor Scheiner, with whom Galileo was on a war footing, were involved in his trial causally and to Galileo's disadvantage, nor was any other Jesuit. No doubt this should be taken seriously.[226] Indeed, while Galileo laments over his opponents, he fails to provide proof of the machinations that he complains about. His source is hearsay. It follows that we should have critical reservations about such statements by Galileo, some of which are very pathetic and emotional. This was gradually to have its effects in the literature about the trial, too. Of course we cannot rule out such oral, private influence on the incident, for which of course, by the nature of the matter, there is scarcely any authentic proof.

It is certain, however, that the Dominicans Caccini and Lorini set in motion the first proceedings in 1616. Whether they can be criticized for this depends entirely on their motives, which however are not known to us. Presumably the close connection that existed from the start between the Thomistically oriented Dominicans and the philosophers of the Aristotelian school was what led to the intervention of the Dominicans. Furthermore, even a critical historian should assume morally inferior motives only when there are at least indications of this that can be taken seriously. But there were also

[225] See Grisar, *Galileistudien*, 321–335; see Müller, *Der Galilei-Prozess*, 173–179; see Soccorsi, "Il Processo di Galileo," 913–918.

[226] William A. Wallace comes to the same conclusion in "Galileo and the Professors of the Collegio Romano at the End of the Sixteenth Century," chapter 2 in Paul Cardinal Poupard, ed., *Galileo Galilei: Toward a Resolution of 350 Years of Debate* (Pittsburgh: Duquesne University Press, 1987).

morally unobjectionable motives for rejecting Galileo. These should not be ruled out in advance, even in the case of the large number of Peripatetics who were exposed to immortal ridicule by Galileo's wit; in Galileo's day they dominated the schools. Perhaps there were such motives even for Professor Cremonini of Padua, who refused to look through Galileo's telescope, since he preferred to rely on Aristotle rather than on his own eyes. Most likely we could assume in their case personal annoyance and the enmity against Galileo that developed from it.

Their hostility toward Galileo, based on the matter itself, which naturally seems absurd to us today, can nevertheless make a claim to be taken seriously as an element in mankind's intellectual struggle to know the truth, just as their adversary Galileo can. Indeed, history must not be written exclusively as the story of the victors and the survivors.

We should not dismiss out of hand the fact that Urban VIII played a decisive role in the 1633 trial. If it had been up to personnel like Fr. Riccardi, Fr. Maculano, Cardinal Francesco Barberini, and others to determine the course of things, the result of the trial would have been a condemnation of the *Dialogo* "*donec corrigatur*" (until it is corrected) and a penance for its author. The fact that that was not the outcome is quite certainly to be attributed to the pope's intervention. Historians have often discussed the remarkable change whereby the former admirer and patron of the Florentine court mathematician now withdrew his favor from him and treated him sternly—though not nearly with the harshness that would have been possible. In this regard it has often been maintained that Urban's friendship with Galileo had turned into enmity after the publication of the *Dialogo*. In fact, the pope must have felt that both his court theologian Riccardi and his private chamberlain Ciampoli had gone behind his back by their collusion with Galileo in issuing the imprimatur for the *Dialogo*.

The decisive factor, however, for the rejection of Galileo's arguments by Urban VIII was a far-reaching philosophical difference of opinion dependent on a theory of science. This had come to light in a conversation with Galileo, the course of which was written down by Cardinal Oregio, the court theologian who was present. During the conversation the pope had asked Galileo whether God could have ordered the course of the heavenly bodies quite differently from the way that Galileo thought he had discovered, without any change in the astronomical phenomenon that we can observe. If Galileo wanted to set up his opinion as absolutely true, then like it or not he had to prove that all other explanations were unthinkable. At that Galileo had to remain silent.[227] The fact that in the *Dialogo* Galileo nevertheless put this argument on the lips of the foolish, stubborn Simplicio may have provoked Urban. The sudden change may also have been caused by the fact that the pope learned about the existence of the special prohibition for Galileo, in other words, the oft-mentioned *registratura* dated February 26, 1616, and therefore had to regard Galileo's conduct as rude disregard for ecclesiastical authority.[228] Moreover there are indications that Urban VIII took very seriously the concern about the integrity of the faith. In the years of his pontificate a series of inquisitorial trials came up; after all, the Inquisition was also responsible, among other things, for combating false prophecies, inauthentic mystical phenomena attributable to magic and superstition, whereby it proceeded without respect for persons.

The uneasiness with which the pope watched over the unity of faith, especially in Italy, caused him to reproach Cardinal Spada, for

[227] On this problem, see Orlando Todisco, "I motivi della condanna di Galileo e la riflessione filosofica di Cartesio," *Sapienza* 36 (1983): 5–19. He cites as his source *A. Oregii S.R.E. Cardinalis Archiepiscopi Beneventani, Philosophicum praeludium: Opus postumum* (Rome, 1687), 119.

[228] See Drake, *Galileo at Work*, 339–340.

instance, with a sharply-worded brief, just because he had risen to his feet when a heretic said grace before a meal. In the year 1630 he demanded that the Duke of Mantua expel the Protestant merchants, because he feared that they would spread their faith in Italy.[229]

Thus Urban plainly feared that the faith was endangered by the Copernican teaching, too, although no one could say for what reasons; indeed, faced with so many friends of Galileo in his immediate circle, he may have felt like the victim of a conspiracy.[230]

Nowadays some point also to a political motivation for Urban's proceedings against Galileo. The author, who relies on the Piccolomini Archive, speaks first about the beautiful, charming scene in the convent in Arcetri, where the mother abbess and the religious sisters embraced Galileo's daughter, Maria Celeste, full of joy that her father had come back home, and shed tears of emotion. Then he continues: "The example of the behavior of the nuns who showed such sympathy to the condemned 'heretic' [!], proves that there are good reasons to judge the trial as a political proceeding against the 'first mathematician and philosopher of the Grand Duke of Tuscany.' "[231] His relationship with the pope deteriorated rapidly during this time, the theory goes, since Tuscany sided with the House of Austria against France, while the pope looked to France as an ally. In addition, Galileo's friend and patron, Archbishop Piccolomini of Siena, was related to the well-known imperial general by the same name. And so our author concludes: "The *raison d'état* [reason of state] of someone like Richelieu, which was superior to man, morality, and science, found a strong echo in the heart of Urban VIII. What depressed the researcher who was absorbed in his science, then, were

[229] See von Pastor, *Geschichte der Päpste*, 610–616.

[230] See M. Viganó, "Contribución al estudio de la 'cuestion galileana,'" *Ateismo e Dialogo* 15 (1980): 145–146.

[231] Šolle, *Neue Gesichtspunkte*, 54.

not the final shadows of a dark night that was passing, but rather the beginnings of the modern era."[232] Such an argument, however, means no more and no less than this: Urban VIII intended to kick the "dog" Galileo as a way to injure the "master" Tuscany.

Another version of the story sees in both cases, the one in 1616 and the one in 1633, chiefly consequences of a feud among Florentine factions, staged in a Roman theater, since those involved were in the main from Florence.[233] Of course it would hardly be possible to produce an authentic proof of this. Moreover, the motives already mentioned are quite enough to explain Urban's severity toward Galileo.

Nevertheless, we cannot speak about Urban's "hostility" toward Galileo, because despite everything the curial officials showed Galileo an altogether extraordinary degree of consideration in comparison to the treatment that was usual then. If Urban had wanted to humiliate or even destroy Galileo, he would have proceeded differently.

But in this context we should consider not only the character and special interests of possible adversaries of Galileo, but also his own personality. "In this drama, over which so much philosophical ink has been spilled, surely the simple fact of personality plays a major role."[234] Galileo's, of course, had a large wingspan. His ingenious intellectual inclinations were combined with a downright Epicurean fondness for good wines and the joys of the table; Galileo's gift for friendship and loyalty, and his caring love for his mother and siblings contrast with the way he treated the mother of

[232] Ibid., 65.

[233] See Giorgio Spini, "The Rationale of Galileo's Religiousness," in *Galileo Reappraised*, 65–66. The author's almost exclusively sociological view of history leads him to this assumption.

[234] Crombie, *History of Science*, 444.

his children. We can list in one column a religious-theological interest and impressively formulated statements of pious, indeed, ecclesiastical sentiments, and in the other a sanguine-choleric temperament which simply could not abide any attempt to call into question his own personality, his own accomplishments. And this is no doubt the dominant impression that the observer of Galileo gets. The way in which he insults and reviles his scientific opponents in a literarily perfect form is just as immoderate as his self-esteem. Even as an aged man with ruined health he still could write sentences like these to his friend Diodati in a letter dated January 2, 1638: "You dear friend and servant Galileo has been completely blind for a month, so that the heavens, the world, the universe, which I expanded a hundred or rather a thousand times more than any philosopher in all past centuries by my wonderful observations and clear argumentation, has now become for me so restricted that it occupies no more space than my own person."[235]

Even before that he had written in his *Saggiatore* to Father Grassi, "What are you going to do, Reverend Grassi, if it was granted to me alone to discover everything new in the heavens and nothing at all to anyone else?"[236]

One author, who elsewhere prosecutes Galileo a bit too harshly, nevertheless accurately writes with regard to such statements, "No one who became acquainted with Kepler's disarming manner could be seriously angry at him. Galileo, in contrast, had a rare talent of stirring up enmities; not the indignation alternating with surprise that Tycho evoked, but rather the cold, relentless hostility that genius plus

[235] Favaro, *Le Opere di G. Galilei,* 17:247; translated from the German version in Müller, *Der Galilei-Prozess,* 157.

[236] Favaro, *Le Opere di G. Galilei,* 6:383; translated from the German version in Müller, *Der Galilei-Prozess,* 40.

presumptuousness minus modesty creates in the circle of mediocrities."[237] These qualities made Galileo despise those who were admonishing him to be cautious and challenging him to seek sound proofs first before he propagated the cosmology of Copernicus.

Without overrating these considerations, they should still be kept in mind if we are to discuss now the objective reasons for Galileo's fate, which are to be sought within the larger framework of science and intellectual history; the effect of the latter reasons, if not induced by the former as by a catalyst, may nevertheless have been accelerated and intensified. First let us recall once again that wide circles of influential theologians and leading ecclesiastical figures had an altogether benevolent and receptive attitude, not only in general toward research in the natural sciences but also in particular toward the Copernican teaching propagated by Galileo.

The fact that Galileo, after his conflict with the Inquisition — and this is true of both cases — was personally on the best of terms with cardinals, bishops, religious, and the popes themselves, demonstrates that it is unlikely that his clash with the Roman officials was motivated primarily by personal resentment. Objective reasons must have led to it.

THE STATE OF SCIENCE AT THAT TIME

We may be able to see the decisive reason for it — after what was probably the official Church position formulated by Bellarmine[238] — in the historical situation of science at that time. In this discussion, of course, there is the danger of an anachronistic judgment. To us today, indeed, it is completely clear that neither Copernicus nor Galileo furnished real proof for the fact that the actual situation in the cosmos

[237] Koestler, *Die Nachtwandler*, 374.

[238] See GT241-242 [*i.e.* the text containing footnotes 59 and 60 in section 2 on "History"].

corresponded to the heliocentric system. No doubt Galileo, above all through his telescopic observations, had provided several weighty if not conclusive reasons for his conviction that the geocentric system could not be true. Even Tycho Brahe already knew that; of course Galileo did not deal with his works. Consequently, though, nothing at all in favor of Copernicus had been proved yet. Our question must therefore be the following: Could this have been realized by Galileo's contemporaries also? And in fact this was the case. Bellarmine as well as Clavius and Grienberger spoke more than once about the necessity of a real proof for the heliocentric system; therefore they had seen through the lack of solidity of the arguments formulated by Galileo. Crombie therefore is correct when he puts it this way:

> Kepler and Galileo both tried to answer the traditional objections against the rotation of the earth, and furthermore to find arguments in favor of the earth's rotation; but because they could see only segments of the complete picture, they could not convince most of their contemporaries.... Proving the earth's rotation remained one of Galileo's main goals in his dynamic studies; nevertheless in the end, despite all his efforts, the only thing that he was able to show was this: the theory that earth rotates is at least just as plausible as the theory that it remains at rest.[239]

Therefore the evidence that Galileo could cite in favor of the heliocentric system in 1633 was nothing less, but also nothing more, than what he had already had in 1616. Namely: (1) the paths of the planets,

[239] Crombie, *History of Science*, 425–426. Compare J. Bernard Cohen, "Newtons Gravitationsgesetz—aus Formeln wird eine Idee," *Spektrum der Wissenschaft* (May 1981): 101–111.

which periodically come closer to the earth and then move away from it again, (2) the sunspots, which show the sun's rotation around itself on its axis, and (3) the tides of the sea.[240] Then, too, the mere application of geometric and kinematic methods was no longer sufficient to decide the question between Tycho Brahe's model and the Copernican one — dynamic considerations would be necessary. Not until 1684, however, did Newton, building on the mechanical-dynamic research from Galileo's final years, discover the laws of gravitation and with their help prove the factuality of the heliocentric system. Another century had to pass before Guglielmini was able to provide the first experimental proof for the movement of the earth.[241]

Only a combination of methods from geometry and physics would have been capable of producing that proof, which Galileo in his time could not yet do at all. Moreover, Galileo, in a scientifically inadmissible way, neglected to come to terms at all with the system of Tycho Brahe, much less to demonstrate its indefensibility.

It is certain, therefore — the two other proofs seem out of the question anyway — that Galileo by no means proved what he challenged his contemporaries to accept.[242] It may be that he himself intuitively knew that the heliocentric cosmology is correct, but he was not capable of proving it conclusively to his scientific

[240] See Mario G. Galli, "L'argomentazione di Galileo dedotta dal fenomeno delle maree," *Angelicum* 60 (1983): 386–427.

[241] See Giorgio Tabarroni, "Giovanni Battista Guglielmini e la prima verifica sperimentale della rotazione terrestre (1790)," *Angelicum* 60 (1983): 462–486. See also below, GT190 f. [in the subsection "New Expert opinions: who issues the *imprimatur*?" after footnote 89.]

[242] In reference to Stillman Drake, Vinaty says that it is wrong to maintain that Galileo had provided no proofs for the movement of the earth, etc., which however is not accurate. Bernard Vinaty, "Galileo e Copernico," in *Galileo Galilei, 350 anni di Storia*, 74. Moreover, Wallace makes an interesting observation in *Collegio Romano*, 96, that in no passage in the *Dialogo* does Galileo make a claim to have proved the Copernican system!

contemporaries and to himself. With regard to the situation in the history of science in Galileo's time we can probably say that, precisely for scientists who thought empirically, the demand to agree with Galileo necessary meant preferring the unripe fruits of an unverified imagination to assured experience.[243] Yet the progress of science lives on the intuition of an ingenious researcher. Intuitions, though, are scarcely communicable; it takes a long time to verify them, as for instance the fate of the theory of relativity shows.

Yet, although skepticism has good reason to oppose intuition, it nevertheless compels a scientist to further intellectual effort, the result of which can then be the transition from intuition and hypothesis to assured knowledge.

Pascal shows what the epistemological situation really was at that time with regard to the various cosmologies, when he writes, "All phenomena of motion and retrograde motion can be derived completely from the hypotheses found in the writings of Ptolemy, Copernicus, Tycho Brahe, and many others. But who will be able to make such a momentous judgment, and who will be able to prefer one at the expense of the others without risking an error!"[244] Then there was the fact that the teaching about the earth's movement contradicted the experience, confirmed anew every day, that it stands fast unshakably, while there was no explanation — in the absence of knowledge about gravitation — for why bodies rest firmly on earth despite its speedy motion through the universe. We must not be surprised, therefore, if physicists and astronomers found themselves contradicting Galileo. Significant examples of

[243] Thus Edwin Arthur Burtt, *The Metaphysical Foundation of Modern Physical Science* (Garden City: Dover Publications, 1955), 38.

[244] Blaise Pascal to P. Noël, Paris, October 29, 1647, in *Oeuvres Complètes de Blaise Pascal*, ed. J. Chevalier (Paris, 1954), 375.

this can be found also in the non-Catholic England of that time.[245] Particularly informative is the position taken by Francis Bacon, who was still one of the protagonists of the experimental methods. In his *Novum Organum* he states that it is inadmissible to accept a movement of the earth.[246]

Therefore it is truly astounding that Cardinal Bellarmine, in a letter dated April 12, 1615, to the Carmelite provincial Foscarini, who was likewise occupied with the question of the day and agreed with Galileo, wrote,

> Thirdly I say, if there really were a proof for it…, then in interpreting the passages of Sacred Scripture that seem to teach the opposite, we would have to exercise the utmost caution and prefer to say that we did not understand them, rather than to declare as false a view which has been proved true. Nonetheless I am of the opinion that there is no such proof, since none has been submitted to me. To demonstrate that the phenomena can be explained correctly, if one assumes that the sun is at the center of the universe, is not the same as demonstrating that the sun is *de facto* in the center and the earth is in the heavens [i.e., space]. I think that in the first case there may be a proof, but I have the greatest doubts concerning the second, and in case of doubt one should not abandon Scripture as it was interpreted by the Fathers of the Church.[247]

245 See Blumenberg, *Die kopernikanische Wende*, 129–131.

246 See Francis Bacon, *Novum Organum Scientiarum* (Leyden, 1650), 310 (Lib. II, Aph. 46).

247 Favaro, *Le Opere di G. Galilei*, 12:171–172.

This position taken by the cardinal therefore corresponds exactly to the state of the scientific knowledge of the time in which this letter was written. As he did already in his statement in 1571,[248] Bellarmine also admits forty-five years later that the proven findings of natural science have a normative function in interpreting the Bible.

What is really scandalous, however, is that, viewed superficially, the scientific question about the structure of the universe could become a theological question in the first place, that anyone could ask the Bible at all about the facts of natural science, and that a natural-scientific statement could become the object of ecclesiastical proceedings. Nevertheless this becomes comprehensible to a great extent against the background of the intellectual history of the Baroque era.

The intellectual-historical situation

In contrast to our world, which is divided up into independent, autonomous fields of inquiry, in which religion, as one area of life among others, struggles laboriously for its right to exist, the man of the Baroque era saw heaven and earth, time and eternity, divine and human affairs, the Church and the world, science, technology, and faith, in one breathtaking, magnificent view as harmonious components of one mighty, all-encompassing cosmos of being that proceeded from God and tended toward God.

Closely connected with this was the extremely noteworthy process of the new discovery of Aristotle throughout the Western world. The result of this, the development of Baroque Scholasticism, took permanent shape with the *Disputationes metaphysicae* by the Spanish Jesuit Francisco Suárez, published in 1597. In that same year the first lecture about Aristotelian metaphysics was given at the Protestant

[248] See Soccorsi, "Il Processo di Galileo," 875, citing an unpublished commentary on Thomas dated 1571.

University of Helmstedt; the *Disputationes* by Suárez, used as a textbook, was the foundation of instruction in philosophy at Protestant institutions of higher learning, too. These facts show that at the turn of the seventeenth century Aristotelian metaphysics exerted a force that could overcome even the walls between religious denominations.[249] Aristotelian-metaphysical or logical thinking was then also what made Bellarmine, the exegete Pereira, and Galileo himself capable of ruling out the possibility of a real contradiction between the truth of revelation and scientific knowledge. But that was not the real reason for the rejection of Galileo by the Aristotelians of his time — not at all. What seemed to them unacceptable was a philosophy, a metaphysics, that was to be measured by the results of natural-scientific and experimental methods.[250] Instead they advocated the stance that the findings of natural science, obtained through observation and experiment, had to present themselves for the judgment of philosophy. What Galileo resisted was this all-encompassing claim of the philosophers — in this case the Aristotelians — and not so much persecution by the theologians.[251]

[249] See Ernst Lewalter, *Spanisch-jesuitische und deutsch-lutherische Metaphysik des 17. Jahrhunderts: Ein Beitrag zur Geschichte iberisch-deutscher Kulturbeziehungen und zur Vorgeschichte des deutschen Idealismus*, Ibero-amerikanische Studien 4 (Hamburg, 1935).

[250] See Jan Campbell, "Some Cultural Problems of the Galilean Period," *Ateismo e Dialogo* 15 (1980): 149–156. See also the pertinent observations by Vasoli, "Tradizione," 204.

[251] See Drake, "Galileo and the Church," 87. The whole article has this tenor. The dominant influence of Aristotelianism disappeared only toward the end of the seventeenth century. See Charles B. Schmitt, "Galileo and the Seventeenth-Century Textbook Tradition," in Galluzzi, *Novità celesti*, 217–228. We might ask, however, whether the boundaries between philosophy and the natural sciences existed or were defined as Drake thinks. See Lisa T. Sarasohn, "French Reaction to the Condemnation of Galileo, 1634–1642," *Catholic Historical Review* 74 (1988): 37.

The intellectual-historical situation of Galileo's time was also determined, not by a totalitarian view, but surely by the idea of the harmony between nature and the supernatural.[252] For a man of the Baroque era, "the world was still entirely contained within the religious realm, and much that seems at first glance to be merely earthly culture ... proves upon closer inspection to be determined by secret connections with the things of the next world. The comprehensibility of the age, even down to the profane areas of economic history and sociology, ultimately lies within the horizon of the religious."[253] If for this reason the documents of the Holy Office severely oppose the teaching that the sun, and not the earth, is the central point at rest, then this may simply be due to the fact, among other things, that those human beings, given their whole mindset, were utterly incapable of *not* thinking theologically, religiously, and consequently it never occurred to them at all to distinguish between salvation history and astronomy. In the universal theological-existential view of Galileo's contemporaries, the earth was indeed first and foremost not a heavenly body that was to be explored according to its physical and astronomical circumstances, but rather the unique arena of divine revelation and redemption. There is a certain charm in calling to mind in this context the experience of the astronauts of our day. Presumably the first men who traveled to the moon could rejoice in their secular success, when their status had changed from drifting in the cosmos to being safe on earth again. For them, viewed existentially, the earth was certainly the center of the cosmos far more than it was for any disciple of Ptolemy.

[252] See Walter Brandmüller, "Sinnenhaftigkeit und Rationalität, Versuch einer geistesgeschichtlichen Bestimmung des Barock," *Internationale Katholische Zeitschrift 'Communio'* 2 (1973): 59–72.

[253] Benno Hubensteiner, *Vom Geist des Barock, Kultur und Frömmigkeit im alten Bayern* (Munich, 1967), 18. See also Johann Auer, *Die Welt — Gottes Schöpfung*, Kleine katholische Dogmatik 3 (Regensburg: Pustet, 1975), 179–191.

The holistic view of the Baroque era was therefore what made the people of that time incapable of separating the realms of nature and revelation, of natural science and theology, when such a separation would have brought clarification.

Therefore we too would be wrong if we were to believe that the Church feared the collapse of the cosmology of antiquity and for that reason rejected Galileo. The Christian faith had overcome similar crises in the past, without experiencing the alarming frame of mind that characterized the intellectual atmosphere of the Galileo trial. They had taken note of the spherical shape of the earth and the existence of antipodes, and they were not frightened by that any more than by Nicholas of Cusa's first hint about the infinity of the universe or by the aforementioned teaching of Nicolas d'Oresme about the movement of the earth. It is quite in keeping with this openness and breadth of theological thought in the Renaissance when for instance Cardinal Cajetan, the important expert and commentator on St. Thomas Aquinas and Luther's conversation partner in Augsburg, in his commentary on Genesis championed the interpretation that the account of the Fall through Original Sin should not be regarded as a historical record but rather as a poetic-dramatic depiction of purely spiritual events.[254]

The opinion that geocentrism "was defended and then lost as a specifically Christian theme"[255] is actually based on a misreading of the intellectual situation of the period between Copernicus and Galileo. "In reality Christianity was not at all embarrassed about a cosmic metaphor for its definition of man's standpoint in the cosmos; a theology *knows* much too much about man, his place and destiny in

[254] See Jacques Marie Vosté, "Cardinalis Cajetanus in Vetus Testamentum, praecipue in Hexaemeron," *Angelicum* 12 (1935): 320–322.

[255] Blumenberg, *Die kopernikanische Wende*, 134.

the world, to have any remaining problematic gaps into which an absolute metaphor could jump."[256]

Denominational-political problems

In the historical moment in which Galileo was brought before the Inquisition, however, people had lost the abovementioned easy impartiality with regard to the letter of the Bible, which had characterized the statements of authoritative theologians; so much so that Galileo's apparent contradiction to the Bible evoked the worst misgivings and opposition among the ecclesiastical authorities. The cause of this was that in Rome the Galileo affair was viewed in connection with the religious and political situation of northern and central Europe, where for a hundred years Protestantism not only had come to rule but also increasingly spread with the help of a corresponding politics. Moreover, in June 1630 the Thirty Years' War had taken an unexpected turn to the detriment of the Catholic powers through the intervention of Gustav Adolf of Sweden. The triumphal march of the Swedish king, the beginnings of which had already made it impossible to carry out the imperial Edict of Restitution in 1629, sped on after the defeat of the Catholic League's commander, the Count of Tilly, near Breitenfeld in September 1631, and reached its climax in the Battle of Rain am Lech in April 1632, in which Tilly met his death. On the eve of the Galileo trial the continued existence of both the Bavarian capital Munich and the Catholic cause in the Holy Roman Empire was directly threatened. Other countries such as Ireland, England, Scotland, Hungary, Austria, Poland, and the Netherlands presented a picture that was no more favorable — not to mention Scandinavia. Thus a look at the denominational map of those days was enough to cause the pope

[256] Ibid.

and the Curia great alarm with regard to the continued existence of Catholic life in Europe.

The prospect of not only dealing with the dangerous religious-political situation but also of running into difficulties through theological disputes within the Church may have put Urban VIII into the defensive posture that made it impossible for him to consider the Galileo affair with the calm and discretion that seems appropriate to us today.

Yet this is understandable, when we remember also that—at least in the view of those who were involved—it was a matter of the integrity and authority of the Bible as the revealed Word of God, that Bible that had become the focal point of the theological disputes with the Reformation.[257] The Catholic Church's theological and religious instinct for self-preservation commanded her to hold fast unshakably to its literal interpretation. For, as the Spanish theologian and bishop Caramuel wrote somewhat later, Where will it lead, if we start to understand the Bible figuratively? In the end, he says, the Protestants would triumph: they understood even Jesus's words at the Last Supper only in a figurative way and therefore denied transubstantiation in the doctrine on the Eucharist. Therefore we ought to be grateful to the cardinals of the Inquisition that with their sentence against Galileo they bolted the door against this false understanding of the Bible.[258]

This leads to a further thought: Protestantism itself had constantly and forcefully emphasized the authority of the Bible as the

[257] "Because of the Protestant revolts, interpretation of Scripture in Galileo's days had become the most sensitive issue there was, and not just any Catholic was free to air his own opinion as obviously true." Drake, "Galileo and the Church," 85. What Drake points out in reply to Galileo's opponent Boscaglia here, of course applies generally.

[258] See Grisar, *Galileistudien*, 270–271.

sole source of faith, as opposed to the Catholic doctrine of the two sources of revelation, the Bible and the Apostolic Tradition. This biblicism took its extreme form in the doctrine of verbal inspiration, a doctrine that maintained that the wording of the Bible, indeed even the Hebrew vowel pointings in the biblical text, can be traced back to divine dictation.[259] Now since the Protestant side again and again leveled the accusation that the Catholic Church had fallen away from the Word of God, the inevitable result was that people on the Catholic side tried to refute this accusation by being as faithful as possible to the wording of the Bible. This was another example of how their imposed defensive posture almost inevitably led to narrower intellectual views that were detrimental to the cause that they had intended to defend thereby. Therefore, primarily it was not at all a question of whether the sun or the earth stands still or moves. In reality it was about the understanding the authority of the Bible and about the dispute with Protestantism. With that we have probably identified the real cause why an ecclesiastical authority suddenly decided to speak out on a question which, as we see today, was purely a matter of natural science, after it had remained silent on it for seventy years.[260]

Form and Content of the Galileo Judgment

Not uncommonly, in statements about the Galileo case, the judgment issued in this matter is described as one of the many errors of the Church's Magisterium, through which the Church disavowed

[259] Pertinent texts are reprinted in Emanuel Hirsch, *Hilfsbuch zum Studium der Dogmatik: Die Dogmatik der Reformatoren und der altevangelischen Lehrer quellenmässig belegt und verdeutscht* (Berlin: De Gruyter, 1964), 314ff., 396.

[260] See ibid., 239–240.

once and for all her claim to infallible teaching authority in matters of faith.[261] But this is not correct.

It should be noted here first that so-called infallible doctrinal decisions can be made only by an ecumenical council or by the pope alone; moreover, as such they are governed by very definite criteria. Only when these are satisfied is there a decision of the highest ecclesiastical Magisterium, and the Church believes that the spokesmen of this supreme teaching authority — pope and council — are preserved by the assistance of the Holy Spirit from error in interpreting divine revelation, which is why the members of the Church are obliged in conscience to the obedience of faith with regard to them.

No such decision was made in the Galileo cases at any stage of the proceedings, if only because the authorities engaged in them in 1616 and in 1633 — the Congregation of the Index and the Inquisition—were not qualified to exercise the infallible teaching authority. It makes no difference that the pope was personally concerned with the affair. It was as clear then to anyone who was theologically informed as it is today.[262]

Naturally this distinction has momentous consequences for the importance of the decrees of 1616 and 1633. If only because they originated with subordinate curial officials, they were as a matter of principle subject to revision, revocable. Numerous statements by contemporaries show that people were aware of this in Galileo's day, too.[263]

But let us take a look now at the content of the decrees.

[261] See Hans Küng, *Infallible? An Inquiry* (New York: Doubleday, 1983 [1971]). See also Joseph F. Costanzo, *The Historical Credibility of Hans Küng* (North Quincy, MA: Christopher Publishing House, 1979), 81–83.

[262] See Appendix III, "L'assenso ai decreti delle SS. Congregazioni," in Soccorsi, "Il Processo di Galileo," 919–926.

[263] See Grisar, *Galileistudien*, 164–170.

The decree by the Congregation of the Index dated March 5, 1616, calls the teaching that the earth can move "*falsam illam doctrinam Pithagoricam, divinaeque Scripturae omnino adversantem,*"[264] (that false Pythagorean doctrine which altogether contradicts Sacred Scripture). This is very significant. Recall that Copernicus's view had been described by the experts of the Holy Office as heretical—in other words, as contrary to the deposit of faith in its binding formulation. The dicastery itself, on the other hand, by no means accepted the interpretation of the experts, but only declared that the teaching in question contradicted Sacred Scripture. Plainly, within the curial offices there must have been considerable oppositions in judging Copernicus. Indeed, the resulting caution of this formulation, *divinaeque Scripturae omnino adversantem* (which altogether contradicts Sacred Scripture) had its consequences! In other words, if it were ever demonstrated that this contradiction with Sacred Scripture only apparently existed, one could at any time cut the ground out from under the congregation's judgment with this demonstration! The formulation of the decree, which obviously relies heavily on the interpretation advocated by Cardinal Bellarmine, therefore also emphasizes thematically the revocable character of the decision.

Exactly the same is true of the sentence from the year 1633. This time, too, the heliocentric cosmology is described as a false opinion contrary to Scripture, but not as heretical. The reason why Galileo was described as strongly suspected of heresy was not his defense of heliocentrism, but rather the fact that he championed an opinion that had already had been condemned in 1616 by the Holy Office as contradicting Sacred Scripture. A judgment of heliocentrism as a

[264] Texts in Pagano and Luciani, *I Documenti del processo di Galileo Galilei*, 102–103.

formal heresy is out of the question.[265] Then too, there was the fact that only the statement "The heliocentric system corresponds to cosmic reality" was rejected. In the form of an astronomical-mathematical working hypothesis, however, it could continue to be advocated, discussed, and developed. Descartes, a contemporary of those decrees, understood the stance of the Inquisition in the same way; he by no means feared impediments to research when he wrote, "If we prove that everything that is derived from the hypothesis agrees with the experiments, then the hypothesis is just as advantageous for life as the knowledge of the truth itself."[266] To summarize: by the nature of things, the path to proving the contrary through further research was deliberately kept open.

But that already raises a further question: the question about the consequence of the judgment against Galileo in intellectual history and in the history of science.

[265] This is certain, contrary to D'Addio, in *DBI* 6 (1964): 109 or 112, who does not take this formal canonical aspect into consideration. See also the very early statement of this argument in Ferraris's *Bibliotheca* (1763), 139. Therefore it is impossible to say that the 1633 judgment condemned the Copernican teaching as heresy! Compare the text of the sentence: *"Ti sei reso . . . veemente sospetto d'eresia, cioè d'aver tenuto e creduto dottrina falsa e contraria alle Sacre e divine Scritture, ch'il Sole."* (Favaro, *Le Opere di G. Galilei*, 19:405). The formula of abjuration, too, speaks about the "*falsa opinione*" of heliocentrism, and about the fact that "*detta dottrina è contraria alla Sacra Scrittura.*" (ibid., 406). The commissioner of the Holy Office, Fr. Maurizio Benedetto Olivieri, O.P., represented the same standpoint in a *votum* dated October/November 1820. See Brandmüller and Greipl, *Copernico, Galilei e la Chiesa*, 375.

[266] René Descartes, *Principia Philosophiae*, pt. 3, no. 44, in *Oeuvres de Descartes* VIII/1, ed. Charles Adam and Paul Tannery (Paris, 1905), 99.

CHAPTER 4
CONSEQUENCES FOR SCIENCE

CONSULTATION OF THE current literature on Galileo from this perspective leads to an almost unanimous judgment: the Church's verdict against Galileo crippled scientific research in Catholic Europe for the future and moreover conjured up the conflict between natural science and the Church, which our time has an urgent duty to overcome. This, as we said, is the tenor of the flood of literature about Galileo since the mid-nineteenth century. In very recent times voices can still be heard which characterize the 1633 judgment as "a total No which entailed the complete alienation of post-Tridentine Catholicism from the progress of the natural sciences and of philosophy, with far-reaching negative consequences especially in Italy."[267] Another contemporary author speaks about a separation — *divorzio* — which for a long time was irreparable between official Catholic culture and the new "thinking."[268]

It is clearly evident, however, that judgments like those cited above can be expressed only by someone who, entangled in invincible prejudging, closes his eyes and ears to a multitude of facts that prove the opposite. Galileo himself — under ecclesiastical supervision, managed as always by the Inquisition — discovered the wobbling of the moon in 1637 as the result of his continued

[267] Eugenio Garin, "Il caso Galileo nella storia della cultura moderna," in Galluzzi, *Novità celesti*, 7–8.

[268] Vasoli, "Tradizione," 73.

astronomical observations, and in the following year he published the *Discorsi* that established his genuine scientific importance.[269]

An unbiased look at Italy's scientific landscape in the seventeenth and eighteenth centuries, moreover, shows us academies that were dedicated to the experimental investigation of nature in all areas, especially in the field of astronomy.[270] In Florence the famous Accademia del Cimento was founded shortly after Galileo's death. Its outstanding promoter and *spiritus rector* was the future Cardinal Leopoldo de' Medici. The Accademia dei Fisiocritici existed then — and exists to this day — in Siena; in Padua, the Accademia Constantium; in Naples the famous Investiganti. Brescia too had its Academy of Natural Sciences. Preeminent among these was the l'Istituto delle Scienze in Bologna, which especially enjoyed papal support. In Rome itself, right under the eyes of the Inquisition, the Accademia Fisico-matematica conducted its work, Magalotti pursued his studies of comets, and Cassini, the discoverer of Saturn's moons, carried out his observations.[271] The telescopes constructed by Ciampini in Rome during these years were in demand throughout

[269] Pages GT110 f. [=the subsection "The Harvest of a Life of Research"]

[270] See the wealth of bibliographical information in Ludwig Hammermayer, "Europaische Akademiebewegung und italienische Aufklärung," *Historisches Jahrbuch der Görres-gesellschaft* 89 (1962): 247–262, esp. 252; Lynn Thorndyke, *A History of Magic and Experimental Science*, VII, VIII (New York: Columbia University Press, 1958); Martha Ornstein, *The Role of Scientific Societies in the Seventeenth Century* (Chicago: University of Chicago Press, 1928); Nicola Badoloni, *Introduzione a Giambattista Vico* (Milan: Feltrinelli, 1961).

[271] Giovanni Domenico Cassini (1625–1712), ancestor of a whole family of astronomers, was among other things close friends with Pope Alexander VII, who pursued astronomy himself. Cassini described and investigated the comets of 1664 and 1665 together with Christina of Sweden, recognized a shadow on Jupiter as its eighth moon, and then found additional ones. A. Ferrari, in *DBI* 21 (1978): 484–487.

Europe,[272] and Alfonso Borelli lectured brilliantly in the learned circles that Christina of Sweden gathered in the halls of what is today the Palazzo Corsini in the Trastevere district. Not only was he known for his investigations of comets, but even before Newton he had stated the crucial conjecture that there must be one principle encompassing the three laws of Kepler.[273]

The fact that even Jesuits were in a lively exchange of ideas with the famous Royal Society of London is a barometer indicating the esteem that Catholic natural scientists enjoyed, even in the England of "no popery."[274]

These simple facts have been known at least since the publication of Hartmann Grisar's *Galileistudien* in the year 1882[275] and Adolf Müller's *Der Galilei-Prozess* in 1909. Hardly a word is spent on those books in the Galileo literature of the following years, which sheds a telling light on it. Prejudices, too, must be tended and passed on!

[272] Giovanni Giustino Ciampini (1633–1698) was a prelate in the Curia from the pontificate of Clement IX on, member of the *Accademia del Cimento*, founder of the *Accademia Fisico-matematica* that was named after him, corresponding member of the *Académie de France*, with international and interdenominational scientific contacts, even with the Royal Society of London. S. Grassi Fiorentini, in *DBI* 25 (1981): 136–143.

[273] Giovanni Alfonso Borelli (1608–1679), vigorously promoted by Cardinal Leopoldo de' Medici yet impoverished during the final years of his life, was a guest in the house of the general superior of the teaching order of "Scolopi," several of whom were his students, who also arranged for the posthumous publication of the works that he left behind. His relevant work was *Theoricae Mediceorum planetarum ex causis physicis deductae* (Florence, 1666). See U. Baldini, in *DBI* 12 (1970): 543–551.

[274] Conor Reilly, "Jesuitica in the Philosophical Transactions," *Archivum Historicum Societatis Jesu* 26 (1957): 339–362; and Conor Reilly "Jesuits and the Royal Society," *The Month* (1957): 108–113. About the successful natural scientific research at the Jesuit University of Ingolstadt in the seventeenth and eighteenth centuries, see Andreas Kraus, in *Handbuch der bayerischen Geschichte*, ed. Max Spindler, vol. 2 (Munich, 1966), 800–804.

[275] Grisar, *Galileistudien*, 337–341.

Therefore it is no longer possible to speak about any stagnation of research in this field in Catholic regions.

Another conclusion may follow from what was just said; namely, that in the century after Galileo's death there was no conflict between the natural sciences and the Church. It was, instead, the breaking apart of the Baroque synthesis of faith and science, world and Church, in the wake of the European Enlightenment that conjured up the conflict between natural science and the Church. In the great encyclopedia of Diderot and d'Alembert we also find a corresponding presentation of the consequences of the Galileo verdict:

> From then on the most enlightened among the philosophers and astronomers of Italy no longer dared to advocate the system of Copernicus, or if they did boldly reveal that they adopted it, they were very careful to note that they regarded it only as a hypothesis and that furthermore they most obediently submitted to the papal decrees concerning this matter. It is to be hoped that a country as rich in intellect and knowledge as Italy might finally be willing after all to admit its error, which is so detrimental to the progress of science, and that people there might think about this subject as we do in France. Such a change would be worthy of the enlightened Pope who rules the Church today, himself a friend of scientists and scholars. It is up to him to issue instructions to that effect to the Inquisition, as he has already done in other less important affairs.[276]

[276] *Encyclopédie ou Dictionnaire raisonné des Sciences des Arts et des Métiers*, vol. 4 (Paris, 1757), 174.

As late as 1774, however, the director of the Royal Bavarian Academy, in a speech about the importance of astronomy, was able to celebrate it as "natural theology" which testifies to "the infinite majesty of God" and leads reason to the "fortress of heaven, to the stars which declare the glory and majesty of the Lord."[277] The gulf widened in the following period to the extent that rationalism and finally materialism attained almost unrestricted dominance over the field of natural science. Now Galileo became in all respects the figurehead of a deliberately atheistic science.

Moreover, our findings are further corroborated by the history that we will now present of the reception of the heliocentric cosmology by seventeenth- and eighteenth-century astronomy, which was by no means effortless or even self-evident.

[277] Andreas Kraus, *Die naturwissenschaftliche Forschung an der Bayerischen Akademie der Wissenschaften im Zeitalter der Aufklärung* (Munich, 1978), 17; 251–252.

Chapter 5
Two Centuries of Debate

There is no need to explain at this point that it is impossible to write an exhaustive history of the reception of the heliocentric cosmology. What we can accomplish here more sensibly, though, is to present a representative spectrum of works and authors that mark the path of the debate about heliocentrism. In doing so, we must take into consideration Galileo's defenders as well as his opponents — and of course the relevant courses of action taken by the Church.[278]

The first objection to Galileo's *Dialogo* came, shortly after it was published, from the Frenchman Claude Guillermet de Beaurégard. In early June 1632, the latter wrote his *Dubitationes* (Doubts), in which he grappled with Galileo's reasoning without setting himself up as a defender of Aristotle or Ptolemy. His accusation against the author of the *Dialogo* was that his interlocutor Simplicius, who defended the Ptolemaic tradition, proved too easy to refute. That is to say, Galileo had not seriously dealt with the Aristotelian-Ptolemaic arguments. It may be, and we can suspect, that by taking this position

[278] A very instructive, albeit incomplete, overview is provided by Dino Cinti, *Biblioteca Galileiana raccolta dal principe Giampaolo Rocco di Torrepadula*, Biblioteca Bibliografica Italiana, Contributi 15 (Florence, 1957). See also Appendix II, "Gli avversari di Galileo," in Soccorsi, "Il Processo di Galileo," 913–918. We should point out the splendidly illustrated, very instructive exhibition catalogue by Luigi Pepe, *Copernico e la questione Copernicana* (Ferrara: Opere della Pubblica Biblioteca di Ferrara, 1993).

Guillermet de Beaurégard intended to improve his chances of succeeding Scipione Chiaramonti in his professorship in Pisa—his hopes were not unfounded, and in 1634 Guillermet de Beaurégard obtained the academic position that he desired.[279]

His predecessor Chiaramonti was in fact probably Galileo's most determined adversary; moreover, before the outbreak of the controversy he had declared Galileo to be *molto intendente delle matematiche* (a great connoisseur of mathematics). As early as 1621 he had given his opinion against Brahe's system in a polemical debate with Orazio Grassi, the Jesuit astronomer and opponent of Galileo, and on this account he had been roundly mocked by Galileo in the *Dialogo.* Now in 1633 Chiaramonti wrote a treatise in which he defended his *Anti tycho*. He dedicated it to Cardinal Francesco Barberini, the nephew of Urban VIII. Already quite advanced in age, this father of twelve children followed in the footsteps of four of his sons and received Holy Orders in 1644.[280] It could be one of history's ironies that his own descendant Barnaba Chiaramonti would later be the one

[279] Claude Guillermet de Beaurégard (Bérigard, Berigardo), born ca. 1590 in Moulins, d. 1663, wrote *Dubitationes in Dialogum Galilaei* (sic) *Galilaei lyncei* (Doubts about the Dialogue by Galileo Galilei of the Lincean Academy) (Florence, 1632). The subtitle reads: *Simplicii vel praevaricatio vel simplicitas, quod nullum efficax superesse Peripateticis argumentum ad terrae immobilitatem probandam tam facile concesserit.* (The prevarication or else the honesty of Simplicius, because he so easily conceded that nothing valid remains of the Peripatetic argument proving the earth's immobility.) See Cinti, *Biblioteca Galileiana,* 186–187; *DBI* 7 (1965): 386–389 (unsigned article); and Giorgio Stabile, "Il primo oppositore del 'Dialogo': Claude Bérigard," in Galluzzi, *Novità celesti,* 277–282.

[280] Scipione Chiaramonti, b. 1565 in Cesena, d. 1652. His first work against Galileo: *Difesa al suo Antiticone e libro delle nuove stelle dalle Oppositioni dell'Autore de'Due massimi Sistemi Tolemaico e Copernicano* (Defense of his *Anti-Tycho* and a book of the new stars from the *Oppositions* by the author of *The Two Major Systems, Ptolemaic and Copernican*) (Florence, 1633). See Cinti, *Biblioteca Galileiana,* 191–193; G. Benzoni, in *DBI* 24 (1980): 541–549.

who — as Pius VII — vindicated Galileo, even though his ancestor had been Galileo's bitter adversary.

Unsurprisingly, after the *Dialogo* was censored several works appeared that set out to refute it by using Aristotelian tools. Among these authors we should mention Giacomo Accarisi,[281] professor of philosophy at the *Sapienza* and a qualificator of the Holy Office, as well as Giovanni Barenghi of Pisa.[282] They were joined by the notable and influential Fortunio Liceti[283] and Giorgio Polacco,[284] originally from Venice, and also by the Jesuit Jacques Grandami.[285] At the same time,

[281] Giacomo Accarisi, b. 1599 in Bologna, d. 1653, had dealt with the material in 1636 in lectures about Aristotle's *De caelo*. They were published under the title *Terrae quies solisque motus, demonstratus primum theologicis, tum pluribus philosophicis rationibus* (The immobility of the earth and the movement of the sun, proved first by theological, then by several philosophical arguments.) (Rome, 1637). See Cinti, *Biblioteca Galileiana,* 204–206; A. Petrucci, in *DBI* 1 (1960): 69–70.

[282] The title of his work reads, *Considerazioni sopra il 'Dialogo' dei due massimi sistemi . . . nelle quali si difende il metodo di Aristotele.* (Considerations about the *Dialogue* of the two major systems . . . in which Aristotle's method is defended) (Pisa, 1638). The work was dedicated to Giovanni de' Medici. See Cinti, *Biblioteca Galileiana,* 209–210.

[283] See Giorgio Abetti, *Amici e nemici di Galileo* (Milan: Bompiani, 1945), 239–251. Liceti's pertinent work: *De lunae suboscura luce prope coniunctiones et in eclipsibus observata* (On the moon's overshadowed light observed at conjunctions and during eclipses). (Udine, 1642). See Cinti, *Biblioteca Galileiana,* 216–219.

[284] His work, *Anticopernicus catholicus, seu de Terrae statione et de Solis motu contra systema Copernicanum, catholicae assertiones* (The Catholic Anti-Copernican, or Catholic statements about the standstill of the Earth and the movement of the Sun, contrary to the Copernican system) (Venice, 1644), was quite weighty. In it he praises Galileo's recantation, with which the astronomer worthily placed himself alongside Augustine and Pius II. See Cinti, *Biblioteca Galileiana,* 223–226; Paschini, *Vita e opere,* 572.

[285] Jacques Grandami, S.J., b. 1588 in Nantes, d. 1672. His work: *Nova demonstratio immobilitatis terrae petita e virtute magnetica* (A new proof of the earth's immobility drawn from its magnetic force) (La Flèche, 1645). See Cinti, *Biblioteca Galileiana,* 227–230; Sommervogel, *Bibliothèque de la Compagnie de Jésus,* vol. 3 (1892), 1668–1670.

the Dutchman and Capuchin Anton Maria Schyrleus de Rheita offered a very critical assessment of Copernicus, Severino, Landsberg, and Tycho, in which he did maintain the immobility of the earth yet developed new hypotheses with regard to the planets. The second volume of his work is characteristic of the holistic worldview of the Baroque era; in it, he presented mystical discourses about the heavenly bodies and their astronomical relations.[286] On the other hand, Andrea Argoli — who wrote almost simultaneously — taught that the earth rotates, although he adhered to the Ptolemaic system.[287]

Although the mention of these names shows how strong the phalanx of Galileo's opponents was in the decade after the publication of the *Dialogo,* the name Bonaventura Cavalieri signals an upheaval that was on the way.[288] Cavalieri came from Milan and joined the Jesuit order at the age of seventeen. Impressed with his outstanding intellectual talent, Cardinal Federico Borromeo recommended him soon afterward to the famous Galileo — exactly one year after his first clash with the Inquisition. A very notable sign of what the cardinal of Milan thought about it! Cavalieri traveled to Pisa, where he studied under Galileo's friend Castelli and enjoyed Galileo's encouragement. In 1629, when he was thirty-one years old, he taught astronomy according to Copernicus and Galileo — unchallenged by the Church — as a professor first in Parma and then in Bologna.

[286] About his work, *Oculus Enoch et Eliae sive Radius Sidereomysticus ... I-II* [The eye of Enoch and Elijah or the Mystical Sidereal Ray ... I-II] (Antwerp, 1645), see Cinti, *Biblioteca Galileiana*, 230–233.

[287] Andrea Argoli, *1570 in Tagliacozzo, †1657. His work: *Pandosion sphaericum* [The Complete Setting of a Sphere] (Padua, 1644). See M. Gliozzi, in *DBI* 4 (1962): 132-134.

[288] Bonaventura Cavalieri, *ca. 1598 in Milan, 1615 entered the Jesuit Order, †1647. See Cinti, *Biblioteca Galileiana*, 310–311; A. Ferrari, in *DBI* 21 (1978): 484 (article about Cassini) and 22 (1979): 654–659.

Thus the two positions stood in opposition to each other, and Galileo's contemporaries had a choice. The example of France shows that this choice was by no means easy. Although it could be assumed that Italian scientists might be more inclined to abide by the Holy Office's decrees of 1616 and 1633, since they lived within its sphere of influence, this was not true at all of the subjects of Gallican France. This makes all the more interesting the position taken by French astronomers and physicists regarding the question of astronomical cosmology.

Research into this matter yields a very revealing fact. Although scholars such as the Belgian Froidmont, as well as Descartes, Mersenne, Pascal, and Gassendi, certainly understood that the Roman decrees had by no means made ultimately binding statements of faith, they could nevertheless not simply pick sides between Galileo and Ptolemy. Galileo's theory still lacked conclusive evidence.[289] Therefore,

[289] See A. Beaulieu, "Les réactions des Savants Français au début du XVIIe siècle devant l'héliocentrisme de Galilée" (Reactions of French experts in the early seventeenth century to Galileo's heliocentrism), in Galluzzi, *Novità celesti*, 373–381. The situation is characterized elsewhere as follows in a way agreeing with the preceding:

> The difficulties . . . in understanding the significance of the new findings emerge in these pages with extraordinary force. Almost always dominant in [these efforts] is the search for some form of compromise between the new theories and the comforting certainties connected with an ancient, very stable tradition. But we should remember that in the [sixteen-]thirties astronomy and cosmology were fields with abundant alternative theories. Bacon had declared his uncertainties about Copernicanism between 1620 and 1623. Between 1625 and 1650 Mersenne, Gassendi, Roberval, and Pascal, as everyone knows, declared that they likewise were unsure.

Citation from L. Borselli, Ch. Poli, and P. Rossi, "Una libera Comunità di Dilettanti a Parigi del 600," in *Cultura Popolare e Cultura Dotta nel Seicento*, Atti del Convegno di Studio di Genova, 23-25 Nov. 1982 (Milan, 1983), 27. See also Descartes's letter to Mersenne, Amsterdam, April 1634, in *Oeuvres de Descartes*, vol. 1, 285, 288.

in a work dedicated to Cardinal Richelieu, Pierre Gassendi presented the three great cosmologies without deciding in favor of one of them.[290] Pascal expressed a similar opinion. He writes, "All phenomena of planetary movements and retrograde motions perfectly agree with the hypotheses of Ptolemy, Tycho, Copernicus, and many others that one could propose, only one of which can be true. But who will dare to make such a major judgment, and who, without risk of error, can uphold one to the detriment of the others."[291]

Similar skepticism, if not outright rejection, was common even in non-Catholic Northern Europe. In his *Novum Organum*, Francis Bacon — at any rate one of the proponents of the experimental sciences — determined that it was inadmissible to assume any movement of the earth.[292] People in Denmark were also critical of the Copernican theory, as the example of Niels Stensen shows.[293]

Things developed quite differently in Italy under the watchful eyes of the Inquisition. Anyone here who dealt with physics or astronomy had received a thorough philosophical education,

[290] Pierre Gassendi (1592–1655). See H. Pohl, in *LThK2* 4 (1960): 525. Gassendi's work: *Institutio astronomica juxta hypotheses tam Veterum quam Copernici et Tychonis* (Astronomical instruction according to the hypotheses both of the Ancients and of Copernicus and Tycho) (Paris, 1647). The book was dedicated to Richelieu. Gassendi and Roberval (1634) both placed all three systems on a par with one another and considered it possible that all three were wrong while the true system was unknown. See Borselli, Poli, and Rossi, "Una libera Comunità," 27. With that, however they had again taken up the position of Thomas Aquinas! See his *Commentaria in libros Aristotelis de caelo et mundo*, lib. II, lect. 17, in *S. Thomae Aquinatis Opera omnia III, ed. iussu impensaque Leonis XIII. P.M.* (Rome, 1886), 186–189.

[291] Pascal to P. Noël, Paris, October 29, 1647, in *Oeuvres Complètes de Blaise Pascal*, ed. J. Chevalier (Paris, 1954), 375.

[292] See Francis Bacon, *Novum Organum Scientiarum* (Leiden, 1650), bk. 2, aphorism 46, p. 310.

[293] See Gustav Scherz, "Niels Stensen und Galileo Galilei," in *Saggi su Galileo Galilei*, ed. Comitato Nazionale per le manifestazioni celebrative del IV. Centenario della nascita di Galileo Galilei (Florence, 1967), 19–31.

especially if he was a cleric. This ensured a keen awareness of considerations from the theory of science. This kind of background alone explains the thesis advanced in Rome in 1638 that the Copernican system should be rejected for reasons from theology and physics, but that the system could not be refuted with astronomical arguments.[294] The leading Jesuit scientist Giovanni Battista Riccioli has special significance in this context.[295] His *Almagestum novum,* published in 1651, is nothing less than the summa of the astronomical knowledge of his time. To this day, these two massive folio volumes are impressive to behold — not least because of the many drawings that they contain.

Regarding cosmology, he essentially followed Tycho Brahe but modified his system at one point that was not insignificant. Although Tycho had taught that the sun and the moon orbited the earth, whereas all other planets orbited the sun, Riccioli advanced the notion that only Mercury, Venus, and Mars orbited the sun while the other planets orbited the earth. As far as his position toward Copernicus is concerned, it should be noted first that he discusses in great detail in his introduction the relevant decrees from the Congregation of the Index and the Holy Office. He correctly emphasizes their binding character, while clearly indicating that they by no means contain definitive statements of faith, since congregations of cardinals are not competent to do that. Therefore it is not heresy at all for someone to

[294] The thesis (in Latin) reads: "The Copernican system, that in fact [!] the earth along with other elements and stars move around the sun, we reject as contrary to the principles of the faith and the reasons of physics, although it is not proved impossible by astronomical reasons" (Favaro, *Le Opere di G. Galilei,* 17:363).

[295] Giovanni Battista Riccioli, S.J. (1598–1671). Biography and bibliography in Sommervogel, *Bibliothèque de la Compagnie de Jésus,* vol. 6 (1895), 1796–1805. His work: *Almagestum novum … I-II* (Bologna, 1651). See Cinti, *Biblioteca Galileiana,* 244–248.

assert what these decrees deny. Conversely, for the same reason it is not an article of faith either that the sun orbits a stationary earth. Nonetheless, a Catholic is obliged to obey these decrees; at the very least, no Catholic should maintain the opposite in an absolute way.

Riccioli holds himself to this standard too, when he joins the discussion about Copernicus. First he establishes that so far the opponents of Copernicus have been unable to upset his system in the view of those who have studied it in greater depth. On the contrary, Copernicus keeps gaining more adherents. Thus the line from Horace aptly describes his system: "It [the Copernican system] draws courage and strength even from the iron that strikes it [the decrees of 1616, etc.] in its misfortune."[296] By taking this position, Riccioli plainly declared to his readers his support for Galileo and Copernicus, without however altering the hypothetical character of their theory. This was not double-dealing and certainly not an expression of hypocritical conformism, but rather shows a keen awareness, based on philosophical-theological education, of the problem from the perspective of the theory of science.

By the year 1656, it was obvious that Galileo was now advancing. Not only could an anonymous treatise refuting the existing arguments against heliocentrism be published in Rome,[297] but two editions of Galileo's works were published as well. First an edition of Galileo's *Trattato della Sfera,* edited by Bernardo Savi and dedicated to Cardinal Giancarlo de'Medici, was published in Rome.[298]

[296] *Carmina,* Lib. 4, ode 4, "Drusus," verses 59–60: "*Per damna, per caedes ab ipso ducit opes animumque ferro,*" cited from *Q. Horatii Flacci Opera,* ed. S. Borzsák (Leipzig, 1984), 109.

[297] The work was entitled, in French: *Mathematical proof of the ineptitudes of Jacques Dubois in his attacks against the hypothesis of Copernicus and Descartes about the earth's movement.* See Thomas Henri Martin, *Galilée, Les droits de la Science et la méthode des sciences physiques* (Paris, 1868), 261–262. Unfortunately the author has not succeeded in finding the treatise.

[298] See Cinti, *Biblioteca Galileiana,* 264–265.

Then in Bologna in the Papal States, and therefore with the approval of the ecclesial censor, a two-volume edition came out, which also included previously unpublished essays.[299] The censored *Dialogo* was not included, but the *Saggiatore* then spiritus rector was. This edition was dedicated to Ferdinand II of Tuscany, and the [patron] of the renowned natural scientific Accademia del Cimento in Florence (the future Cardinal Leopoldo de' Medici) contributed decisively to its production. The editor Carlo Manolessi spared no effort to enhance Galileo's reputation. Not only did he preface the work with a very fine portrait of Galileo by the famous engraver Francesco Villamena, but he also reprinted an admiring letter written by then-Cardinal Maffeo Barberini — soon to be Pope Urban VIII — addressed to Galileo, and a poem that he had composed in Galileo's honor. But both the dedication and the preface paid very special tribute to Galileo's scientific greatness in an elevated, panegyrical tone, concluding with the assertion that "*l'opere del nostro gran Galileo, ad onta de'malvagi persecutori sono per accrescersi fama sempre mai più gloriosa, e per servire di fida scorta, e di chiarissima fama*" (the works of our great Galileo, to the shame of the wicked persecutors, are to accrue for themselves an

[299] The imprimatur had been granted both by the archbishop of Bologna, Girolamo Boncompagni, and also by the local inquisitor. The title: *Opere di Galileo Galilei, Linceo Nobile Fiorentino, già Lettore delle Matematiche nelle Università di Pisa, e di Padova, di poi Sopraordinario nello Studio di Pisa, Primario Filosofo, e Matematico del Serenissimo Gran Duca di Toscana* (Works of Galileo Galilei, a noble Florentine member of the Lincean Academy, former lecturer in mathematics at the Universities of Pisa and Padua, then Honorary Professor at the *Studio* in Pisa, Principal Philosopher and Mathematician of the Most Serene Grand Duke of Tuscany.) Collected in this new edition, and supplemented by several treatises by the same Author that are no longer in print. To the Most Serene Ferdinand II, Grand Duke of Tuscany) (Bologna, 1656). The second volume (by mistake?) notes the publication date "1655." Volume 1 had 806 pages; volume 2 had 792. See Cinti, *Biblioteca Galileiana*, 259–264.

ever more glorious fame, and to serve as a faithful guide, and a most distinguished renown).[300]

Although the *Dialogo* was missing, it still meant a great deal that the collected works of an author who had already been disciplined could be published under these circumstances. Galileo's star was in the ascendant. Of course, that does not mean that skepticism and even protest had vanished. Evidence of this is a statement by the Jesuit Honoré Fabri, who after fourteen years of teaching philosophy in Lyon, where he also dedicated himself to the natural sciences and mathematics, became the penitentiary at St. Peter's Basilica in 1660. A statement is attributed to him to the effect that the Bible passages cited against Copernicus should be taken literally until the contrary is proved. Previously none of the leading adherents of Copernicus had dared to make this claim.[301]

Meanwhile, the scales tipped more in favor of Copernicus and Galileo. The Florentine Accademia del Cimento played a not insignificant

[300] See also the beginning of the foreword: "The fame of *Signore* Galileo has now become as great as his genius, since both of them, after having passed beyond the limits of our inhabitable world, have gone on to make their memories immortal in the eternity of the stars."

[301] Honoré Fabri, S.J. (1607–1688). See B. Schneider, in *LThK2* 3 (1959): 1333; Sommervogel, *Bibliothèque de la Compagnie de Jésus*, vol. 3 (1892), 511–521. Reportedly he wrote to a Copernican as follows:

> Some of you, including some of your leaders, have been asked more than once whether they have any proof of the earth's movement; they have never dared to claim this; therefore there is no reason why the Church should not understand those passages [i.e., from Sacred Scripture] in the literal sense and declare that they should be understood that way, as long as there is no proof to the contrary; if perhaps someday you devise one (which I hardly believe), then in that case the Church will in no way hesitate to declare that those passages should be understood in the figurative and improper sense, as in the verse from the Poet: *Terraeque urbesque recedunt* [And the countries and the cities depart].

Quoted, in Latin, from A. Auzout, *Lettre à Monsieur l'Abbé Charles* (Paris, 1665), 49.

role in this. It was their noble patron Leopoldo de' Medici who emphatically supported the great Giovanni Alfonso Borelli, principally by having an astronomical observatory built for him in San Miniato. Among the many fruits of his studies was the very interesting hypothesis of a force of attraction between satellites and their planet on the one hand and between planets and the sun on the other, which is counterbalanced by the corresponding centrifugal forces. This hypothesis extended the debate about astronomy while making very critical use of all the preliminary work that had been done by Kepler, Brahe, Galileo, and others. Naturally, he maintained that the earth moved around the sun. Obviously, this meant that he had to grapple with Riccioli as well and try to refute his arguments for the immobility of the earth.

He found an influential comrade-in-arms for this enterprise in the Jesuit Stefano degli Angeli, Galileo's successor in the chair of mathematics in Padua. Just like Galileo in his *Dialogo,* degli Angeli utilized a sleight of hand in his argumentation by saying that he intended only to investigate what *would* happen *if* the earth was in motion — which of course he did not believe. In his *Considerazioni,* degli Angeli reached an important milestone in the history of Copernicanism.[302] The Jesuit mathematician Antonio Baldigiani probably summed up the scientific development to date when he wrote in 1678 to Galileo's student and biographer Vincenzo Viviani: "*Non si condanna più Galileo per le sue dottrine, né si dice pure che sia eresia contro la Scrittura, di dubbia fede, ma solo si disputa*

[302] Stefano degli Angeli, b. 1623 in Venice, d. 1697 as Galileo's successor in the professorship in Padua. At first a Jesuit, after the suppression of the order he became a secular priest in 1668. His chief work: *Considerazioni sopra la forza di alcune ragioni fisico matematiche addotte dal M. R. P. Giovanni Battista Riccioli ... contro il Sistema Copernicano espresse in due dialoghi* (Considerations about the validity of several arguments from physics and mathematics adopted by G. B. R. ... against the Copernican system, expressed in two dialogues) (Venice, 1667). See M Gliozzi, in *DBI* 3 (1961): 204–206.

sul modo con cui scrisse, che è questione molto diversa dalla prima" (Galileo is no longer condemned for his teachings, nor is it said that they contain any heresy against Scripture or doubt about the faith; the only dispute is about the way in which he wrote, which is a very different question).[303] His confrere Adam Kochański — who had met Viviani during his time teaching at the Jesuit college in Florence and had kept up an exchange of scientific ideas with him — made more or less the same point in an essay that he wrote in 1685, interestingly enough for the *Acta Eruditorum* published in Leipzig.[304] While he too regarded the Copernican system as a still-unproven hypothesis, he nevertheless furnished several observations that could speak in favor of this system, and furthermore indicated that the efforts to provide the relevant proofs are legitimate. In the event of that the Copernican system was proved, the Bible passages often cited in this context could be explained in such a way that a contradiction would be ruled out — the long-standing position taken by Bellarmine. Thus every theological and biblical impediment to research was removed once again.

Preparing the way for Galileo's triumph

At the close of the seventeenth century, the questions raised by Copernicus and Galileo were therefore completely open to debate for astronomers and physicists, though that debate had been defused

[303] Quoted from D'Addio, in *DBI* 6 (1964): 115. About Antonio Baldigiani, S.J., b. 1647, d. 1711 in Rome, see C. Sommervogel, *Bibliothèque de la Compagnie de Jésus*, vol. 2 (1890), 828; and vol. 8 (1898), 1732–1733.

[304] See A. A. Kochański, "Considerationes Physico-Mathematicae circa diurnam Telluris vertiginem," in *Acta Eruditorum* (1685), 317–327. Adam Adamandy Kochański, S.J., b. 1631 in Dobrzyń, d. 1700. See D'Addio, in *DBI* 6 (1964): 115; Sommervogel, *Bibliothèque de la Compagnie de Jésus*, vol. 4 (1893), 1139–1141. About Bellarmine's understanding, see D'Addio, in *DBI* 6 (1964): 35–36.

theologically. It remained to be seen what effect Newton's findings would have.

The year 1710 brought with it a sensational event for the Church: the first edition of Galileo's *Dialogo* with an ecclesiastical imprimatur since it had been censored in 1634. What made it even more sensational was that the *Dialogo* was reprinted together with Foscarini's short work that had likewise been on the Index and with excerpts from the no-less-prohibited works of Zúñiga and Kepler. Probably as a result of an order by the official censor, the judgment against Galileo and the wording of his recantation were also included in this edition.[305]

The same thing was done with Galileo's *Letter to the Grand Duchess Christina,* probably for the purpose of weakening the biblical objections that had been raised in the past—something that Galileo had succeeded quite well in doing with this document, which the Inquisition had described as orthodox during his lifetime.[306] With this edition, all the decrees ever issued by the Congregation of the Index in this matter had been brushed aside—without any backlash from the Church. Basically, the Galileo controversy was *via facti* settled.

Eight years later, a three-volume complete edition of Galileo's works was published without the *Dialogo,* which was available in a

[305] See Cinti, *Biblioteca Galileiana,* 319–320. Unfortunately no trace whatsoever of a pertinent procedure could be found in the Archive of the Holy Office. This edition is not mentioned at all by J. H. Zedler, ed. *Grosses vollständiges Universal-Lexicon aller Wissenschaften und Künste* (Halle-Leipzig, 1732–1750), vol. 10 (1735), 133–136, or by J. S. Ersch and J. G. Gruber, eds., *Allgemeine Enzyklopädie der Wissenschaften und Künste* (Leipzig, 1818–1882), I/52 (1851), 380–395.

[306] The preface gives the following motives for it (in Italian): "in which [Letter to Madame Christina] a response is given with very solid reasons, drawn from the most widely approved Church Fathers, to the calumnies of those who with all their might strive not only to banish his opinion about the constitution of the Universe, but also to bring perpetual infamy upon him personally." See Cinti, *Biblioteca Galileiana,* 319–320.

separate edition. This edition was published in Florence as well, although it included nothing more than a very general notice *Con licenza de'Superiori* (with the permission of the superiors). Whether this was intended to also serve as an ecclesiastical imprimatur is unclear, though it is quite likely.[307]

Ten years later, James Bradley made his sensational discovery of the stellar aberration of the light from the fixed stars and thereby made possible something close to a proof for the motion of the earth. Shortly thereafter his work was translated into Italian (1734) and circulated freely throughout Italy. This was another significant milestone along the victorious path of Copernicanism.[308] At this time, the outstanding Jesuit Boscovich proved to be a follower of Copernicus — even though he still complied with the existing state of the law by writing in one of his works that it was not permitted there in Rome to teach that the earth moves.[309]

[307] Again the foreword is indicative, where it says (in Italian) that Galileo "attained such a high and lofty *segno* [sign or standard], which others had never reached, and by beautifying the sky with new light and new stars and by adorning it with the wondrous tidings of discoveries, and by soaring in a happy, vigorous flight above the other flights, and by making observation, the glorious inventor of new sciences, judge of the truth, he discovered stupendous novelties in the heaven and in nature . . . which had all been hidden and concealed from antiquity." The three volumes contained 718, 722, and 542 pages respectively. See also Cinti, *Biblioteca Galileiana*, 322–328.

[308] See D'Addio, in *DBI* 6 (1964): 116.

[309] See R. G. Boscovich, *De Aestu Maris Dissertatio* (On the head of the ocean) (Rome, undated [1747]), 4. Ruggero Giuseppe Boscovich (Rudjer Josip Bošković, S.J.), b. 1711 in Dubrovnik, d. 1787. His father was Serbian; his mother was from Bergamo. Even before his priestly ordination (1744) he was an internationally recognized scholar, a member of the Arcadia, a poet, and an extremely productive scientific author. In 1761, as a Jesuit, he was elected a fellow of the Royal Society in London. His Copernican-Galilean position on the question of cosmology could be gathered from his works: e.g., Christian Maire and R. G. Boscovich, *De litteraria expeditione per pontificiam ditionem ad dimetiendos duos meridiani*

The reason why he explicitly spoke only about Rome may have been that a new edition of Galileo's works had been published in Padua three years before. His formulation therefore insinuates that this was not a universal prohibition. As a matter of fact, the aforementioned Padua edition of Galileo's collected works marked a new stage in the ongoing and by no means untroubled process of Galileo's reception.

To this day, no one has really explored what motives and intentions prompted the very active and well-known publishing house of the Paduan major seminary to think around the year 1740 about producing a new and comparatively expanded edition of Galileo's works.[310] Perhaps it was partly because the seminary's library was in possession of a copy of the *Dialogo* that contained Galileo's own handwritten notes. In any case, these marginalia were included in the new edition.[311] This edition is particularly significant because it was published with the explicit permission of the Holy Office. The editor was the Paduan priest Giuseppe

gradus et corrigendam mappam geographicam … suscepta (A literary expedition through the Papal States to survey two southern positions and to correct the geographical map) (Rome, 1755), 385–393; and, even earlier, R. G. Boscovich, *De observationibus astronomicis et quo pertingat earundem certitudo* (On astronomical observations and to what extent they are certain) (Rome, 1742), esp. 22–24. See also P. Casini, in *DBI* 13 (1971): 221–230; and Sommervogel, *Bibliothèque de la Compagnie de Jésus*, vol. 1 (1890), 1828–1849. Recently Boscovich's work has been the subject of in-depth research at the University of Augsburg under the direction of Prof. Dr. Klaus Mainzer. See also *The Philosophy of Science of R. Boscovich, S.J., Proceedings of the Symposium of the Institute of Philosophy and Theology* (Zagreb, 1987).

310 See Cinti, *Biblioteca Galileiana*, 335–339. The volumes have 601, 564, 490, and 352 pages respectively. Volume 4 contains the *Dialogo*.

311 See Claudio Bellinati, "Il 'Dialogo' con postille autografe di Galileo, Precisazioni sul Cod. 352 del Seminario di Padova," in Galluzzi, *Novità celesti*, 127–128; Reusch, *Der Index der verbotenen Bücher*, II/1, 399–400.

Toaldo, then just twenty-five years old, who evidently as early as 1741 had approached the inquisitor with jurisdiction over Padua on this account.[312]

The latter then had recourse to the Holy Office in a query dated September 29, 1741.[313] Plainly, Benedict XIV wanted the matter to be resolved quickly and affirmatively, since he delegated special authority to handle the request to the Assembly of Consultors — the *Consulta* — for their *Feria Secunda* (Monday meeting) on October 9, 1741, instead of to the cardinal-inquisitors who were away from Rome. They subsequently informed the inquisitor in Padua that he should allow the *Dialogo* to be printed under the conditions that he himself had recommended.[314] What happened next is difficult to piece together. All we know for sure is that the cardinals, despite the decision that had already been made, also dealt with the proposal from Padua themselves, plainly giving it a great deal of attention; not until their *Feria Quarta* (Wednesday meeting) on June 13, 1742, did they came to the decision that the publication of the *Dialogo* should proceed and that it should include both the Inquisition's judgment dated June 22, 1633, and also Galileo's recantation.[315]

Meanwhile, the print run was underway and volume four, which contained the *Dialogo,* now also included the 1633 judgment against Galileo and his recantation. This was clearly modeled on the 1710

[312] See Marco Restiglian, "Nota su Giuseppe Toaldo e l'edizione toaldiana del Dialogo di Galileo," *Studia Patavina* 29 (1982): 723–724.

[313] See S. C. Aso, *Indicis Acta et Decreta,* 1734–1746, fol. 341; Brandmüller and Greipl, *Copernico, Galilei e la Chiesa,* 151. On this topic see M. B. Olivieri, *Di Copernico e di Galileo,* 96.

[314] See ibid.

[315] See S. C. Aso, *Collectanea,* Index under "Galilei": Censura librorum, 1742–1743, no. 6.

edition of the *Dialogo,* although the additional texts included in the later edition were not printed then.[316]

Thus it was possible for the first complete edition of Galileo's works — one that did not leave out the *Dialogo* — to see the light of day with the ecclesiastical imprimatur. It is no mean feat that the *Dialogo* was reprinted together with the aforementioned texts. By putting assertion and counter-assertion side by side, as though equally legitimate, the actual state of the discussion could be documented. Indeed, one could still hear voices that were very skeptical of the heliocentric system or rejected it outright.

One such voice was that of Eusebius Amort, a canon at Polling in Bavaria and a scholar who was well-known far beyond the borders of Bavaria and even in Italy.[317] In his *Philosophia Pollingana,* which was

[316] Worth noting, however, was Toaldo's foreword, in which he writes (in Italian), among other things:

> As for the main Inquiry into the movement of the Earth, we too comply with the Author's retraction and declaration, declaring most solemnly that it neither can nor should be admitted except as a pure Mathematical Hypothesis, which serves to explain certain phenomena more effectively. Therefore we have elevated, or reduced to Hypothetical form the marginal notes, which were not or do not appear quite indeterminate; and for the same reason we added the Dissertation by P. Calmet, which explains the meaning of the passages from Sacred Scripture pertaining to this matter according to the common Catholic belief. Nevertheless, the *Dialogo* appears in its entirety; except that in several places for greater clarity an addition was made that the Author himself wrote on one of his printed copies, which is preserved in this Seminary Library. These additions were printed in a different type as proof of the good faith in which we are acting. Concerning these we again repeat the abovementioned declaration that we do not wish to depart in the least from the venerable prescriptions of the Holy Roman Church.

Opere di G. Galilei, vol. 4 (Padua, 1744), "A chi legge" ("To the Reader"), no pagination.

[317] Eusebius Amort, Canon Regular (1692–1755), one of the most important theologians of the eighteenth century. See O. Schaffner, in *LThK2*

published in 1730 in Augsburg, he presented a wide-ranging and detailed refutation of the system of Copernicus in order to develop his own cosmology. The fact that they were operating an observatory in Polling and highly appreciated the natural sciences prevents us from rejecting Amort's discourse out of hand as the product of narrow-minded thinking. Even in Protestant circles a decision in favor Galileo had not yet been made; this fact is demonstrated by the widely known astronomical charts *Atlas Coelestis* and *Atlas Novus Coelestis* by Johann Gabriel Doppelmayr — the Nuremberg astronomer who had been educated in Altdorf, Halle, and Leyden. Both works, which were published in Nuremberg in 1742, presented the three great cosmologies of Ptolemy, Copernicus, and Brahe side by side.[318]

Later on, too, there was still opposition to Galileo's heliocentrism. It came from the philologist, philosopher, and pedagogue Gregorio Bressani, who as a priest and theologian had come to a newfound appreciation of Plato and Aristotle via the mathematics and physics of his time.[319] The candor of his thinking and of his character is shown by the close friendship he maintained with the radical Enlightenment thinker Francesco Algarotti,[320] who was ten years his junior and who held him in such high esteem that he introduced him

1 (1957): 446–447; C. Toussaint, in *DThC* 1 (1903): 1115–1117. His work: *Philosophia Pollingana ad normam Burgundicae* (Augsburg, 1730).

318 Johann Gabriel Doppelmayr (1677–1750). See Eckhard Pohl, "500 Jahre Astronomie in Franken," *Jahrbuch des Historischen Vereins für Mittelfranken* 95 (1990/1991): 99–100.

319 Gregorio Bressani, b. 1703 in Treviso, d. 1771. See U. Baldini, in *DBI* 14 (1972): 196–197. The works pertinent here: *Il modo di filosofare indrodotto dal Galilei ragguagliato al Saggio di Platone e di Aristotele* (The philosophical method introduced by Galilei compared to the Essays of Plato and Aristotle) (Padua, 1753); and *Discorsi sopra le obiezioni fatte dal Galileo alla dottrina di Aristotele* (Discourses on Galileo's objections to Aristotle's teaching) (Padua, 1760). See Cinti, *Biblioteca Galileiana*, 340–341.

320 Francesco Algarotti (1712–1764). See E. Bonora, in *DBI* 2 (no year): 356–360.

at the court of Frederick II in Berlin. Presumably, conversations with Algarotti, who was a follower of Galileo, led Bressani to reflections of an epistemological sort that caused him to have considerable doubts about the importance of merely empirical findings. This finally resulted also in his opposition to Galileo.

The new Index

Now, of course, the scales were beginning to tip more and more in favor of Galileo and the heliocentric system. When it came time to produce a new edition of the *Index librorum prohibitorum* during the pontificate of Benedict XIV, the Church authorities also faced the question of whether they should uphold the general prohibition of books teaching heliocentrism contained in previous versions of the Index. As a result of deliberations, the course of which can no longer be reconstructed for lack of sources, the Congregation of the Index decided on April 16, 1757, after consulting with the pope, not to include this prohibition in the new edition of the Index. When we inquire about the specific reason for this renewed intra-curial discussion of heliocentrism, we run into an interesting chronological coincidence. At the same time that the Index was being revised, the Congregation of the Index was also examining volume 4 of the great *Encyclopédie*, published in 1754.[321] As we saw above, in this volume the author of the article "Copernicus" had taken up a position on the Church's prohibition of his work and had demanded the

[321] See *Encyclopédie ou Dictionnaire raisonné des Sciences, des Arts et des Métiers*, vol. 4 (Paris, 1754). In the congregation on May 10, 1757, in which the cardinals discussed the new Index, Abbot Orsini, librarian of the Conti Family, gave a presentation about volume 4 of the *Encyclopédie*. The cardinals decided to make a judgment only after all the volumes were available. Thus the affair was postponed for several years. See also S. C. Aso, *Indicis Acta et Decreta*, 1749–1763, 129.

abolition of the rule requiring that the heliocentric worldview be presented only as a hypothesis.[322]

Since the author had made a direct appeal to Benedict XIV in his article, it is surely not misguided to assume that the pope, who was highly esteemed for his erudition and known as a promoter of the natural sciences, wanted to heed this appeal. So it happened that on April 16, 1757—by whatever method—the decision of the Congregation of the Index was made: "*Quod habito verbo cum Sanctissimo Domino Nostro omittatur decretum quo prohibentur libri omnes docentes immobilitatem solis et mobilitatem terrae,*" (that after consultation with His Holiness, the decree prohibiting all books that teach the immobility of the sun and the movement of the earth should be omitted).[323]

With these words, the prohibition on teaching the heliocentric system was suppressed as soon as the new Index had been put into effect by a papal bull dated December 23, 1757, and published shortly thereafter.

Scarcely a decade after this decision, the French astronomer Joseph Jérôme Lalande took a trip through Italy and also published a report of his travels.[324] In his remarks about his stay in Rome, he also

[322] *Encyclopédie,* vol. 4, 173–174.

[323] Favaro, *Le Opere di G. Galilei,* 19:419, citing S. C. Aso, *Indicis Acts et Decreta,* 1749–1763, 129. The statement by Bernard Jacqueline that in 1664 the general prohibition of books teaching Copernicanism was replaced by a prohibition of books teaching heliocentrism and thus the name *Copernicus* was stricken from the general part of the Index, seems to the author, after inspecting all editions of the Index since 1616, to be unfounded. The name *Copernicus* never appears in the general part of the Index. Jacqueline, "La Chiesa e Galileo nel secolo dell'Illuminismo," in *Galileo Galilei, 350 anni di Storia (1633–1983): Studi e richerche,* ed. P. Poubard (Rome, 1984), 192, Concerning the 1758 Index, see Reusch, *Der Index der verbotenen Bücher,* II/1, 38–41, which however is not about the present question.

[324] See J. J. Lalande, *Voyage d'un françois en Italie fait dans les années 1765/66,* vol. 5 (Venice-Paris, 1769). About Joseph Jérôme Lalande (1732–1807), see L.-G. Michaud, *DBI* 22 (no year): 603–613.

mentions the curial congregations there, including the Congregation of the Index and, of course, the *Index librorum prohibitorum*. In this connection, he expresses his astonishment that the list includes works such as those of Copernicus and Boerhaave, which to him seemed far from any suspicion of heresy. Yet he does not criticize this at all. Rather, he admits that hypotheses in physics and astronomy — and he does classify heliocentrism as a hypothesis — may contain elements that appear dangerous in their remote consequences, which would be reason enough to put them on the Index. He also mentions the decision of 1758 to lift the general ban on heliocentric books, noting that it took quite a lot of effort by scholars before the heliocentric system, which was so well substantiated, found favor with the Congregation of the Index.[325]

However, Lalande was not satisfied with this. He took the opportunity of his stay in Rome to present to the prefect of the Congregation of the Index[326] his wish that the specific listing of Galileo's *Dialogo* on the Index be suppressed.

Cardinal Galli replied that his request was still opposed by a decree of the Holy Office, which required amendment before any further judgment could be made. Lalande also had the impression that Clement XIII was favorable to his request. But Lalande himself did not have the time to visit the many personages whom he would have had to win over for his plan.[327]

To all appearances the astronomer was still as preoccupied as ever with the question of the Church's stance on the Copernican

[325] See Lalande, *Voyage*, 48–49.

[326] Antonio Andrea Galli (1697–1767), created cardinal in 1753, major penitentiary and prefect of the Congregation of the Index. See R. Ritzler and P. Sefrin, *Hierarchia catholica medii et recentioris aevi*, vol. 6 (Padua, 1958), 17.

[327] See J. J. Lalande, *Astronomie* I (Paris, 1771), 540–541.

system. In his *Abrégé d'Astronomie* published in 1775, he discussed the biblical objections to Copernicus, emphasizing those which plainly were still felt to be unresolved.[328] It is a question of the biblical manner of expression, which could be no different if it was to be understood at all; besides, nothing is said in any of the oft-cited Bible passages that touches on Church dogma or physics.

Furthermore, Lalande wrote, several theologians have furnished a wide variety of arguments for the idea that all passages referring to the motion of the sun could likewise be applied to the motion of the earth without doing violence to the texts.[329] It would be a sign of a downright ludicrous zeal to claim that Holy Scripture is entirely free from colloquial forms of expression. Now since Rome had stricken heliocentric books from the Index, there was hope that soon complete freedom would be granted to physicists in this regard, even more explicitly than before.

The question now is whether and, if so, to what extent this view of things was received by theologians. The paradigm for this reception may be found in the oft-reprinted *Prompta Bibliotheca*

[328] "But when they are read without prejudice, we see in them an ordinary language which could not be any different without becoming unintelligible, and we see in them nothing that appears to pertain to dogma or physics. Moreover several ecclesiastical authors have collected arguments of every sort in support of the opinion that the various Scripture passages that speak about the sun's movement can be understood about the movement of the earth without doing violence to them. It would be a very odd zeal that claimed to exclude from the Holy Books all expressions that are common in society [i.e., colloquial]. Besides, the court of Rome no longer has scruples in this regard. They even removed from the last edition of the Index the article that concerned all the books in which the author supports the movement of the earth, and when I was in Rome I saw that there was reason to hope that soon they would more explicitly restore to physicists complete liberty in this regard." J. J. Lalande, *Abrégé d'Astronomie* (Paris, 1775), 173 (translated from French).

[329] Lalande unfortunately does not say which authors he means here.

Canonica... by the Franciscan Lucius Ferraris, a "Church lexicon" that was first published in 1746 in Bologna. It was widely circulated and endured well beyond its own century.[330] Ferraris deals with the present question in two places in his work, most extensively in the article "Mundus" and then under the heading "Haereticus."

Of the twenty-four folio pages of the first article, he devotes less than two to the major cosmological systems.[331] He begins with a presentation of Ptolemy, then describes Copernicus, and finally Tycho Brahe. As far as Copernicus is concerned, after a brief biographical note about the canon of Frauenburg Cathedral he mentions his predecessors from antiquity and then lists his modern-day followers. Among them he names Kepler, Galileo, Landsberg, Boulliaud, Descartes, and Newton. His assessment is as follows: the Copernican system — *aliis omissis rationibus* (leaving aside any other reasons) — plainly contradicts Sacred Scripture and must be completely rejected *tamquam thesis*. He goes on to cite the usual biblical passages and concludes from them that the views held by Copernicus, Pythagoras, and Galileo were rightly condemned as presumptuous and heretical in 1633 under Urban VIII because they contradict Holy Scripture. Therefore, if anyone were

[330] Lucius Ferraris, O.F.M. (1687,–1763). See Z. da San Mauro, in *EC* 5 (1950): 1195. The title of his work varies: *[Prompta] Bibliotheca canonica, juridica, moralis, theologica ... in octo tomos distributa*. After two printings were published in Venice — the second of which contained an appendix with additions by Ferraris — the first Roman edition was published in 1759–1761, which not only inserted the additions from the second printing into the text but also included the numerous additions by another unnamed author, as well as a subject index. These alterations — with regard to the present topic of discussion — were incisive corrections, to which Ferraris issued responses in the 1763 Bologna edition. Here we quote from the first Roman edition and the Bologna edition of 1759 or 1763.

[331] See Lucius Ferraris, *[Prompta] Bibliotheca canonica, juridica, moralis, theologica ... in octo tomos distributa* (Rome, 1759–1761 or Bologna, 1763), 5:177–179.

to assert that the earth moves and the sun stands still, he would be a heretic, since the Holy Office defined the opposite. He then proceeds to refute the Copernican system, once again concluding that it must not be maintained as a thesis. However, this system can be proposed as a hypothesis, as the Holy Office itself allowed in 1620. After all, to propose a hypothesis is not to assert that something is actually the case. Rather, it merely describes something as possible.[332] This is followed by a brief presentation of the Tychonic system and a reference to the countless other cosmological systems. But he lists these three so that everyone can make his own choice — though he is sure that no one, however knowledgeable he may be, is capable of comprehending how wondrously the universe was ordered by God. Ferraris observes that Paulus Rubeus rightly says in his remarks on a decision by the Roman Rota[333] that astronomers and astrologers who speak so confidently about the planets could well be asked to answer the question put by Diogenes: Pray tell how many days it has been since you came down from heaven? For as the saying goes, "*Quantum metitur Astronomus, tantum mentitur Astrologus*" (As much as the astronomer measures, so much the astrologer lies).

Several things stand out in these remarks. First, Ferraris does not correctly report the actual events of 1616 and 1633.[334] Plainly he was

[332] "But that is the distinction between a Thesis and a Hypothesis: because a Thesis asserts and concludes that the thing is so and truly exists, and defends it as such; whereas a Hypothesis asserts nothing altogether certain as truly and actually existing, but only deduces from a thing which is constituted in a certain way that something is possible, but it does not determine whether the thing is that way" (ibid., 178, translated from Latin).

[333] See P. Rubeus (Rossi), *S. Rotae Romanae Decisionum recentiorum* 9/2 (Rome, 1661), 20.

[334] See Ferraris, *[Prompta] Bibliotheca*, 5:161–164.

misled by his informants without having examined the matter thoroughly.[335]

Second, he remarks explicitly that he is making only biblical arguments *aliis omissis rationibus*, which probably reveals a certain intentional ambiguity. Meanwhile, he also misrepresents the legal situation: heresy was never mentioned in any decision by a curial dicastery, and the Holy Office never defined the opposite of the Copernican theory as dogma — which it could not have done, in any case.[336]

So it is not surprising that in the first Roman edition of the *Bibliotheca*, which was published between 1759 and 1761, annotations from the pen of a well-informed critic — presumably that of the Theologian of the Pontifical Household — were added in this passage, as in many others, while the author was still alive. As far as the present case is

[335] He cites G. B. Riccioli, *Almagestum novum...*, vol. 2 (Bologna, 1651), 495–499 (lib. 9, sect. 4, cap. 40); Fortunatus a Brixia, *Philosophia sensuum mechanica ad usus academicos accomodata*, vol. 2 (Brixiae, 1736), 74–75 (tract. 1, diss. 2, sect. 4, propos. 3): "Therefore the Copernican hypothesis about the earth's movement and the sun's immobility plainly contradicts Sacred Scripture, which explicitly teaches the opposite, and therefore legitimately and deservedly was proscribed in 1616, during the reign of Paul V, and in 1633, during the reign of Urban VIII, as rash and heretical"; D. Ursaya a Bosco, *Institutiones criminales...* (Rome, 1701), 19 (lib. I, tit. VI, no. 2): "The dogma is said to be false in the position of a genus; for there are many false assertions, and dogmas about natural things, for example about the number of the Elements, about the shape and size of the Earth, about the order and movement of the Heavens and the Stars, which nevertheless are *not heresies*"; N. A. Caferri, *Synthema Vetustatis sive Flores Historiarum ab orbe condito ex illustrium scriptorum monumentis ... excerpti* (Rome, 1670), 178: "In the year 1633 (on June 22), the opinion of Galileo Galilei, a Most Excellent Mathematician, and Most Keen-sighted Florentine observer of the Stars, formerly [the teaching] of Copernicus and of Pythagoras, that the Sun is the center of the Universe and immobile, while the earth circles it by a daily movement, is condemned by a sentence issued by the Cardinals of the Holy Roman Inquisition." The three quotations are translated from Latin.

[336] See Ferraris 127.

concerned, this annotation is found in the article "Haereticus."[337] It is not without a note of sarcasm: it calls Ferraris a *hospes omnino in re theologica* (an utter stranger in theological matters), only to emphasize then quite correctly that by no means can we speak about a condemnation of the Copernican system as heresy. Even though the qualificators of the Holy Office under Paul V may have passed this judgment, their ruling was never accepted and pronounced by the Holy See itself. Even the suspicion of heresy that Galileo incurred, which led to his condemnation and recantation in 1633, did not refer thematically to the Copernican system as such, but rather to the fact that he stubbornly defended a position that the Holy Office had censored for being contrary to Scripture. Furthermore, it is clear that if the philosophers (i.e., the natural scientists) were to prove with evidence that Copernicus was right, it would immediately be permissible to accept his theory. In that case, one would be able to say that the language of the Sacred Scripture refers to visible celestial phenomena, not to their underlying physical or astronomical causes. But an even sharper rebuff is dealt to Ferraris in an annotation to the article "Mundus."[338]

Having been attacked in this way, Ferraris defended himself at the next opportunity, which was provided by the Bolognese edition of his work. In it, Ferraris marshaled the same authorities he had cited once before, but added another to the list who had fallen into the same error as Ferraris. Once again, however, he did not examine the matter thoroughly.[339]

[337] See Ferraris, *[Prompta] Bibliotheca*, vol. 3 (Rome, 1759), 336–337; Ferraris, *[Prompta] Bibliotheca*, vol. 3 (Rome, 1767), 305 (note on "Haereticus").

[338] See Ferraris, *[Prompta] Bibliotheca*, vol. 9 (*Additiones et Supplementa*) (Bologna, 1763), 140; Ferraris, *[Prompta] Bibliotheca*, vol. 5 (Rome, 1760), 212.

[339] See A. M. de Lugo, *Dizionario storico portatile* (Naples, 1754), article "Galileo." D'Addio, in *DBI* 6 (1964): 116, thinks he found in this annotation a (hypothetical) decree of the Holy Office from 1712, which of course cannot be verified elsewhere. It is probably not the case. On the

However, an investigation of Ferraris's *Bibliotheca* shows the poor quality of information available to both Ferraris and his sources, as well as their carelessness. Even the anonymous Roman commentator deserves this reproach, because even he was not aware that the Index of 1758 no longer contained the relevant book ban.

However, what remains valuable about the remarks by Ferraris is his skepticism at the level of the philosophy of science, which made him insist on the term "hypothesis": *Quandonam de coelis venistis?* (When did you come down from the skies?) This question, addressed to the overconfident astronomers, was still justified even after 1758. The removal of Copernican books from the Index did not signify the Church's approval of the Copernican system; it only meant that biblical or theological misgivings were no longer obstacles to accepting it.

It was in keeping with this "new" state of affairs that no one raised objections when Giovanni Battista Guglielmini published in 1789 an initial experimental proof for the rotation of the earth — and did so in Rome with the ecclesiastical imprimatur.[340]

Skirmishes while in retreat

In the same year that Guglielmini's Latin version appeared in Bologna, the ex-Jesuit Girolamo Tiraboschi,[341] the successor to Muratori and already renowned during his lifetime, delivered a lecture at the

---

one hand, nothing about this text suggests that it is a quotation. Grammar alone forbids this assumption. On the other hand, the statement here very probably refers to the 1620 decree from the Congregation of the Index.

340 See G. B. Guglielmini, *Riflessioni sopra un nuovo esperimento in prova del diurno moto della terra* (Rome, 1789); and Guglielmini, *De diurno terrae motu experimentis physico-mathematicis confirmato* (Bologna, 1792). On this subject see Giorgio Tabarroni, "Giovanni Battista Guglielmini e la prima verifica sperimentale della rotazione terrestre," *Angelicum* 60 (1983): 462–486.

341 Girolamo Tiraboschi, S.J. (1731–1794). See Sommervogel, *Bibliothèque de la Compagnie de Jésus*, vol. 8 (1898), 34–48. We quote from the reprint in G. Tiraboschi, *Storia della Letteratura Italiana*, vol. 10 (Rome, 1797), 362–387.

Accademia de'Dissonanti in Modena. In it, he presented not only an astoundingly accurate history of early Copernicanism in Italy, but also an account of the Galileo case that was not without critical undertones.[342] Still, he made it perfectly clear that the Church had never condemned the followers of the Copernican system as heretics, and that the overly strict measures taken against Galileo were measures of the Holy Office, to which even the most zealous Catholics never attributed the prerogative of infallibility. Indeed, this entire affair shows how wonderful God's providence is for not allowing the Church to issue a formal doctrinal judgment in this matter, although the majority of theologians were convinced that Copernicus contradicted the Bible.

Nonetheless, Tiraboschi opines, the handling of the Galileo case was felicitous in every respect, since there was too much reliance on the philosophers of the Peripatetic school who — for their part helpless against Galileo's arguments — had entrenched themselves behind Holy Scripture. They should have dealt more seriously with the problem of biblical hermeneutics.

So much for Tiraboschi, who can probably be viewed as a representative of the judgment of the academic world of his day.

Tommaso Mamachi, O.P.[343] — the Theologian of the Pontifical Household at the time—took offense at these remarks, however, and his comments were added to the 1797 Roman edition of Tiraboschi's work. Mamachi, who enjoyed an excellent reputation as an historian and archaeologist, disagreed now with Tiraboschi's

[342] In doing so he relied on A. H. J. F. De Bérault-Bercastel, *Histoire de l'Église,* 24 vols. (Paris, 1778–1790), 21:140–146.

[343] Tommaso Maria Mamachi, O.P., b. 1713 on the Island of Chios, d. 1792, appointed 1781 Master of the Sacred Palace. See V. M. Fasola, in *LThK2* 6 (1961): 1338; Gaetano Moroni, *Dizionario di erudizione storico-ecclesiastica* 103 volumes (Venice, 1840–1861), 42:95–97. Anfossi relates that the anonymous annotations to Tiraboschi's work originated with him. See Maffei, *Giuseppe Settele,* 461.

assertion that the Copernican system was *evidentemente confermato e dimostrato,* and noted that even many followers of Copernicus did not view it as proven. Tiraboschi expressed his opinion about this in a "*Lettera al rev. Mo Padre N. N...*"[344] that he also reprinted in his

[344] See Tiraboschi, *Storia della Letteratura Italiana,* 388–389. Here is the letter, translated from the Italian original:

> If it was right not to allow in this Capital of the Catholic World the *History of Italian Literature* by Abbot Girolamo Tiraboschi to be reprinted without the addition of several annotations that served as an antidote to the insufficiently moderate statements in it, and to the passages that are insulting to several Supreme Pontiffs, which can be read scattered here and there in it; then there is even more reason not to give permission to reprint without any corrections that letter in which all those statements and those summarized passages are collected, along with attempts to justify them by discrediting as useless or inconclusive the annotations added to them. Hence to prevent the prejudice that could result from reading it for those who either did not take the trouble to compare thoughtfully the passages of the *History* with the notes made on them, or else were not capable of spotting the tricks with which the Author of the Letter tries to overcome his Adversary, at the foot of the page a brief analysis will be made of the annotations which call for examination, and their appropriateness and validity will be demonstrated.
>
> Since the *History* says that the Copernican system was conclusively confirmed and proved in our times with many arguments and experiments, and citing a Letter by Galileo in which he seeks to explain the passages from Sacred Scripture that oppose it, and observing that the same Galileo was condemned neither by the Universal Church nor by Rome, but only by the Tribunal of the Inquisition, some Readers of the same *History* could be induced to think that the aforesaid system was properly demonstrated, and that it would be licit to support it as a thesis; and that the biblical passages that depict the earth as immobile did not have to be understood literally, and to promote the idea that our Historian was also of this opinion. It was therefore helpful to warn them [1] that the aforementioned system was not considered proven, not even by many who adopted it; [2] that since it was not proven, it was not necessary to abandon

1797 edition. In this rather satirical, exaggeratedly courteous letter, Tiraboschi admits that one can in fact speak about "proof" only when there is a strict, geometrical demonstration. A second comment by Mamachi likewise spurred Tiraboschi to reply: he expressed his profound gratitude to the Reverend Father for having

---

the literal sense of the Sacred Scriptures, which depict for us the daily and also yearly movement of the sun; [3] that the reasons adduced to prove the earth's movement should be rejected as nonexistent and fallacious, according to the rules bequeathed to us by the Fathers of the Church, particularly by St. Augustine, and [4] that because of the veneration and obedience that the Author of the *History* professed to the Holy Apostolic See, it was considered that he, in reporting historically what he had read in several books about the Copernican system, had never thought to oppose the rulings of the Supreme Pontiffs Paul V and Urban VIII, the former by a Decree of the Sacred Congregation of the Index dated March 6, 1616. It suppressed and prohibited books that defended the mobility of the earth and the stability of the sun, and through the Venerable Cardinal Bellarmine and the Commissioner of the Holy Office it admonished and commanded Galileo not to maintain the aforesaid opinion or to teach it verbally or in writing, because it contradicted the Sacred Scriptures; the latter approved the sentence passed on the same Galileo by the Tribunal of the Holy Roman Inquisition, and the condemnation of the *Dialogue* by him about the two Major Systems of the World, the Ptolemaic and the Copernican, which was prohibited by the aforesaid Sacred Congregation of the Index on August 23, 1634. Hence it cannot be said [1] that Galileo was condemned only by the Tribunal of the Inquisition. [2] That if the Author of the Annotations interpreted too favorably the Historian's intention, he is not blameworthy, but praiseworthy for having judged according to the Christian maxims of charity; [3] that the prohibition issued by the Sacred Congregations of the Holy Office and of the Index against maintaining the Copernican System as a Thesis was acknowledged as reasonable even by the famous Mathematician Christian Wolff, although he is a professed Lutheran, in his *Discursus preliminaris philosophicae rationalis* [Preliminary Discourse on Rational Philosophy (Verona, 1735), 65ff.].

expressed his conviction that he, Tiraboschi, never intended to contradict the decrees of Paul V and Urban VIII concerning Copernicus and Galileo. Furthermore, he pledged never again to use the inappropriate expressions reproved by Mamachi.

Now when this "Letter to the Most Reverend Father N. N." was presented to Mamachi together with the edition of the *Storia della Letteratura Italiana* to obtain permission to print, he did not hesitate to include comments about this, too. These account for most of the objections raised in 1820 by Mamachi's successor against Settele's *Elementi di Ottica e di Astronomia,* which are nearly identical to those from 1797.[345]

Incidentally, this controversy shows how the relation between freedom of scientific debate and ecclesial censorship was understood in Rome at that time. It played out at the same time that the astronomer Giuseppe Calandrelli in his Roman observatory was observing and describing the orbits of Mercury (1786). Finally, thanks to the financial support of Pius VII, he was able to publish, unchallenged by any censorship, his *Opuscoli,* which took the Copernican system as their starting point.[346]

This was increasingly assumed as self-evident, and when the Dominican D. Pini published his exactly eighty-four-page pamphlet *L'incredibilita del moto della Terra brevemente esposta* (A brief explanation of why the earth's movement is incredible) in Milan in 1806, it only elicited laughter.[347] At the same time and in the same city, work was underway on the massive thirteen-volume edition of Galileo's

[345] Tiraboschi, *Storia della Letteratura Italiana,* 389–390 See C. Wolff, *Philosophia rationalis sive logica* (Frankfurt-Leipzig, 1729), *passim.*

[346] Giuseppe Calandrelli, b. 1749 in Zagarolo (Rome), d. 1827, was highly esteemed internationally as a mathematician, physicist, and astronomer. See U. Baldini, in *DBI* 16 (1973): 440–442. His work, which is also cited by Settele: *Opuscoli Astonomici I-VIII* (Rome, 1803–1824).

[347] See Cinti, *Biblioteca Galileiana,* 349–350.

works, whose eleventh and twelfth volumes contained the *Dialogo* along with his recantation and the 1633 sentence. This was meant to highlight the national glory of Italy, which surpassed all other nations in the arts and sciences.[348]

This then was scientific situation on the eve of the Settele affair, which we are about to describe: the Copernican system — corrected, substantiated, and refined by the results of nearly two hundred years of astronomical research — had met with ever greater approval, or rather, it had become increasingly self-evident, even though — at least judging by today's standards — Friedrich Wilhelm Bessel was the first to provide conclusive proof for this system with his discoveries in the year 1838.

It would not be wrong to see this as a consequence of the fact that wide swaths of the educated class had increasingly adopted the mindset of the empirical natural sciences, while philosophical reflection about its presuppositions and import was at home now only in "conservative" circles. The result was an optimism about human knowledge that could become positively intoxicated by empirical results, which were now multiplying explosively, and in the process could forget the epistemological problems and ethical dilemmas associated with them.

[348] See ibid., 351–356.

Part Two

The Solution

Chapter 1
The Context of the Proceedings in 1820

The investigation of the significant events in intellectual and cultural history surrounding the imprimatur for the astronomical textbook by Professor Settele, who taught the heliocentric cosmology, leads us to Rome in the years around 1820, which are generally included in the "Restoration" epoch. These were the final years of Pius VII (Barnaba Chiaramonti), who was elected in 1800 and died in 1823.[349]

The parameters for the lives and experiences of that generation were and remained decisively determined by the influence of the man after whom the "Napoleonic era" was named. The French Revolution and its consequences were the upheaval that broke up the governmental system of Europe in many, far-reaching ways, particularly in Italy. The collapse of the Papal States in Napoleon's brutal clutches, accompanied by sickening outrages against the Church, remained engraved indelibly in the memory of his contemporaries. Inasmuch as they were members of the Roman Curia, they could not forget those years, in which they were scattered to the four winds and their dicasteries were de facto dissolved and incapable of functioning.

[349] Josef Schmidlin, *Papstgeschichte der neuesten Zeit*, vol. 1 (Munich, 1933), 1–366.

Who among his contemporaries could have been able to blot out from his memory the indescribable mistreatment of eighty-year-old Pius VI by Napoleon, who abducted the elderly pope in 1798 and first carried him off to various localities in Italy and finally to Valence, France, where Pius VI passed away on August 29, 1799?

Because of the very difficult circumstances, a conclave could not be held until March 14, 1800. The new election took place in Venice, thanks to the discretion and energy of the future Cardinal Ercole Consalvi, with the support of the Viennese court, from which Barnaba Chiaramonti emerged as Pius VII.

Although it was possible for him to restore papal rule in Rome, nevertheless the very next year the events surrounding the conclusion of the concordat with Napoleon boded ill for the future. The rude reception given to the pope who, under pressure from Napoleon, had traveled to Paris, only to find himself there again in the non-speaking role of an extra at Napoleon's coronation as emperor of the French in 1804, continued in a series of further arbitrary Napoleonic actions, which reached their climax in the occupation of Rome by French troops (February 1808) and the annexation of the Papal States by France (May 17, 1809).

Thereupon Pius VII publicly excommunicated Napoleon, and then the same fate befell him as his predecessor's: he too was violently taken, first to Grenoble and finally to Savona, while the Cardinal Secretary of State Pacca was imprisoned in the fortress of Fenestrelle. New afflictions awaited the pope and the likewise deported cardinals, when they were commanded to participate in the non-canonical wedding of Napoleon with Marie Louise of Austria (April 2, 1810); most of them refused and stayed away. For this reason, Napoleon forbade them to wear the purple, and so those cardinals, sixteen in all, headed by Consalvi, went down in the history of resistance to the Corsican as the "black cardinals."

After that followed, in 1812, the deportation of Pius VII to Fontainebleau and that cynical coup by Napoleon with which he obtained the signature of the physically broken pope on agreements that were damaging to the Church's interests, which the emperor immediately proclaimed as a new concordat. For the pope and his loyal assistants there was no forgetting the days of severe pangs of conscience that prompted Pius VII to revoke his signature.

Then, when Napoleon's rule broke down and the Papal States were restored by the Congress of Vienna in 1815, Pius VII and Consalvi, now secretary of state again, could think about rebuilding from the ruins.

The year was now 1820. What the authoritative figures had experienced and suffered in the last two decades was ever-present to them in a scarcely undiminished form; it had become part of their life and had left its mark on their personality and thinking. Under these circumstances it would have been all too understandable if the papal and curial circles had developed a decidedly reactionary mindset, nostalgic for the *"Ancien Régime"* and hostile to "progress."

Although this may be more or less true about the political-administrative-economic government of the Papal States — Consalvi's reforms were quite remarkable — nevertheless this hardly applies, without further qualification, to what we would call today scientific, educational, and cultural policies. Admittedly, until now this field has not been adequately considered as an object of investigation, not even by scholarship on the history of the city of Rome; nevertheless, the little that is known allows us to conclude that research, writing, and teaching thrived in Rome in the post-Napoleonic years. No doubt the great financial difficulties that burdened the Papal State as a result of the political upheavals were an almost insurmountable obstacle to governmental promotion of science.

The Lincean Academy,[350] and also the Arcadia[351] and the Archeological Academy,[352] held their meetings. Most importantly, however, the Accademia di Religione Cattolica proved to be very lively.[353] After the interruption from 1809 to 1814, it experienced a meteoric rise. It recognized that its real mission was to come to terms with contemporary science, which was influenced by the Enlightenment, and the academy fulfilled this task with numerous lectures. On the eve of the events that will be presented here, the *Giornale Arcadico* was also founded by the then-thirty-year-old Prince Pietro Odescalchi.[354] His program extended far beyond the originally literary-poetic purposes of the famous Arcadia, in that through this newly founded periodical he wanted to inform the interested public about the findings of research in the natural sciences. It served as a literary forum for archeology, too, which by the way was especially promoted by Pius II. Increasing numbers of foreign travelers streamed to Rome, especially from the countries of Northern Europe. German, English, and Scandinavian artists and literary figures rendezvoused on the banks of the Tiber. Moreover, in Rome they were regarded as welcome guests. They contributed significantly to the exchange of information and ideas.

The Roman universities had much to suffer, both from the political upheavals and also from the oppressive public and private

[350] See Enrica Schettini Piazza, *Bibliografia storica dell'Accademia Nazionale dei Lincei* (Florence: Olschki Editore, 1980).

[351] See the exhibition catalogue *Tre secoli di storia dell'Arcadia* (Rome, 1991), which also contains several essays on the topic. See also Maria Teresa Acquaro Graziosi, *L'Arcadia, Trecento anni di storia* (Rome, 1991).

[352] See Carlo Pietrangeli, *La Pontificia Accademia Romana di Archeologia* (Rome, 1983).

[353] See Antonio Piolanti, *L'Accademia di Religione Cattolica, Profilo della sua Storia et del Tomismo*, Biblioteca per la storia del Tomismo, 9 (Vatican City: LEV, 1977).

[354] See GT214 f. [= approx. 2nd page of section 6, "causa finita"]

poverty. The Sapienza of Pius VII, which was reopened immediately after his election, had been enriched by the same pope, especially with professorships in the natural sciences and medicine. A delicate problem had arisen: the majority of the professors had taken the oath of loyalty to the regime of the French occupiers. Pius VII solved it generously: those who were most compromised were only obliged to attend three days of spiritual exercises.[355] A few years after the pope's return, the university had five faculties with forty-three professorships. The fact that Pius VII established a professorship for *"fisica sacra"* on the faculty of philosophy of the Sapienza testifies to the Holy Father's sensitivity to the intellectual challenges of his time. It was expected that the appointee would grapple with the anti-religious interpretation of the findings of research in the natural sciences.[356] A glance at the program of lectures of the Accademia di Religione Cattolica shows that the latter had the same purpose.[357]

Since 1816 Cardinals Litta, Fontana, di Pietro, della Somaglia, and Pacca had been working with the secretary of their commission, Msgr. Bertazzoli, on a reform of the universities for the entirety of the Papal States. These are names that we will encounter again repeatedly.[358]

[355] See A. P. Bidolli, "Contributo alla storia dell'Università degli Studi di Roma, La Sapienza durante la Restaurazione," *Annali della Scuola Speciale per gli Archivisti e Bibliotecari dell'Università di Roma* 19/20 (1979/1980): 76. An overview can be found in Nicola Spano, *L'Università di Roma* (Rome, 1935), 61–69.

[356] See A. Piolanti, *L'Accademia di Religione Cattolica, passim.*

[357] See Bidolli, "Contributo," 76.

[358] See Schmidlin, *Papstgeschichte*, 173–175; Massimiliano Pavan, "Le istituzioni culturali nella Roma Napoleonica," *Rivista Italiana di Studi Napoleonicai* 24 (1987): 119–135; Bidolli, "Contributo," *passim*. The latter also cites the older literature.

Chapter 2
The Congregation of the Index and the Holy Office

Dicasteries and procedures

The work by Nicholas Copernicus, *De revolutionibus orbium coelestium* (On the revolutions of the heavenly bodies), published in 1543 in Nuremberg, was placed on the Index of Forbidden Books in 1616 together with the likewise heliocentric works by Zúñiga[359] and Foscarini,[360] and also, later on, by Kepler.[361] After that, only the Sacred Congregation of the Index[362] was competent to modify or

359 See Didacus a Stunica (Diego López de Zúñiga) in *Job Commentaria* (Toledo, 1584). The page in question is numbered 205.

360 See Paolo Antonio Foscarini, *Sopra l'opinione de'Pittagorici e del Copernico nella quale si accordano e si appaciano i luoghi della Sacra Scrittura e le ripropo-sizioni teologiche che possono addursi contro di tale opinione* (Naples, 1615).

361 See Johannes Kepler, *Epitome Astronomiae Copernicanae* I (Linz on the Danube, 1618).

362 See Reusch, *Der Index der verbotenen Bücher*, 429–434; Johann Heinrich Bangen, *Die Römische Curie, ihre gegenwärtige Zusammensetzung und ihr Geschäftsgang, nach mehrjähriger eigener Anschauung dargestellt* (Münster: Aschendorff, 1854), 124–145; Georg Phillips, *Kirchenrecht*, vol. 6 (Regensburg: Verlag von Georg Joseph Manz, 1864), 598–624; Niccolò Del Re, *La Curia Romana, Lineamenti storico-giuridici*, Sussidi Eruditi 23 (Rome: LEV, 1970), 325–329; Moroni, *Dizionario*, vol. 16 (1842), 211–216; vol. 34 (1845), 166–173. The most thorough discussion is Jacques Marie Joseph Bailles, *La Congrégation de l'Index mieux connue et vengée* (Paris: Veuve Poussielgue et fils, 1866).

even revoke this measure. Moreover, the Holy Office, too — or to use its correct title, *Sacra Congregazione della romana ed universale Inquisizione* — had already been involved in the procedures in 1616. Moreover, Galileo's trial had taken place before its tribunal in 1633 and had ended with his sentencing and abjuration. This also explains the jurisdiction of the Holy Office in the subsequent fortunes of this cause célèbre.[363] With the Master of the Sacred Palace,[364] another entity comes into play. These three authorities should therefore be introduced before we discuss the events concerning Copernicus and Galileo in the 1820s.

We should also note at the outset that their collaboration, as far as the two congregations are concerned, can no longer be reconstructed in detail, due to the unsatisfactory archival tradition.[365] Therefore we have to ask whether these contacts between their officials often took place orally and quite informally. After all, the two dicasteries were housed in the same Palazzo Santo Uffizio, and their personnel were in not a few cases identical. Even the Master of the

[363] See Reusch, *Der Index der verbotenen Bücher*, 1:169–179; Del Re, *Curia*, 89–101; Phillips, *Kirchenrecht*, 6:583–598; Bangen, *Die Römische Curie*, 91–124; Moroni, *Dizionario*, 16: 220–237; vol. 36 (1846), 40–46; Camillo Henner, *Beiträge zur Organisation et Competenz der päpstlichen Ketzergerichte* (Leipzig: Duncker und Humblot, 1890), 367–374; Paul Hinschius, *System des katholischen Kirchenrechts mit besonderer Rücksicht auf Deutschland*, vol. 1, Das Kirchenrecht der Katholiken und Protestanten in Deutschland (Berlin: De Gruyter, 1869), 448–451.

[364] See Moroni, *Dizionario*, vol. 41 (1846), 199–219; Phillips, *Kirchenrecht*, 6:541–545; K. Schrödl, in *KL2* 8 (1893): 463–465. A classic monograph is Joseph Catalanus, *De Magistro Sacri Palatii, Apostolici libri duo* (Rome, 1751).

[365] It should be noted here that, in contrast to the excellent presentation of the history of the Congregation of the Propaganda by Josef Metzler and his colleagues — *Sacrae Congregationis de Propaganda Fide Memoria Rerum 1622–1972*, vols. 1–3 (Rome, Freiburg, Vienna, 1971–1976) — there are no historical descriptions of the dicasteries of the Office and the Index. An urgent desideratum!

Sacred Palace was *ex officio* a member of the Holy Office and of the Congregation of the Index.

Of the three aforementioned dicasteries, we will introduce the Holy Office first. Its prefect, until the reorganization of the Curia by Paul VI, was at any given time the pope himself; this was supposed to express the preeminent importance of the purity of the faith and thus also the rank of the congregation. Therefore the congregation was also called "*Suprema Congregatio*."

The cardinal who actually directed the dicastery therefore bore the name "Secretarius." The first of the officials assisting him was the "commissarius Sancti Officii" (commissioner of the Holy Office) — Bangen and Phillips say so, at any rate — whose duty it was to conduct trials, whereby he was supported by two "socii" or "compagni" (associates). Both the commissioner and the associates traditionally belonged to the Dominican order, specifically to the Lombardy Province of the order, in keeping with a privilege granted by Pius V.

The commissioner was assisted by the "assessor Sancti Officii," who belonged to the secular clergy and assumed a privileged rank among the prelates of the Curia. Since it was his duty to report at the plenary sessions of the cardinals of the congregation, the assessor was usually more prominent than the commissioner. The sequence in the *Diario di Roma* suggests an inverted ranking: the assessor before the commissioner! Next to him in rank were the "promotor fiscalis" and the "advocatus reorum," whose functions corresponded approximately to those of a district attorney and counsel for the defense in the secular judicial system. These were the higher-ranking permanent personnel of the dicastery.

The actual congregation, however, of which the abovementioned officials were the executive agent, consisted of a number of cardinals — eight or nine, later even more. To them belonged the jurisdiction that they exercised in common. Their decisions in each case were submitted to the pope and confirmed by him, usually *vivae vocis oraculo* (verbally, not necessarily in writing).

In order to reach its decision, the congregation employed consultors. These were usually bishops active in other curial functions, too, although the majority of them were high-ranking religious residing in Rome, who not infrequently also taught as professors at Roman universities. The general of the Dominican order, the Master of the Sacred Palace (who will be discussed separately later), and a Conventual Franciscan were "consultores nati" (ex officio consultors) of the Holy Office. These were reminders of the importance of these two orders for the beginnings of the Inquisition in the thirteenth century. A separate group was formed by the so-called qualificators, whose name is derived from their task of determining more precisely the truth or error contained in the teachings or writings that were submitted to them. Among them for the most part were highly regarded theologians and canonists.

The congregation, which had been erected by Paul III in 1542, received this particular organizational form during the reign of Sixtus V by the authority of his bull *Immensa* (1587), with which this pope thoroughly reorganized the Roman Curia.[366] It still existed by and large in the year 1820, in which the events to be depicted here occurred. This bull also established the procedures that are essentially observed to this day. According to them, the consultors met every Monday and submitted the results of their deliberations to the cardinals, who at their Wednesday meeting made a decision on the matter. Accordingly, to this day the latter meetings are simply called "Feria Quarta" (Wednesday). Meetings in the pope's absence took place on Thursdays.

[366] See *Bullarum ... Romanorum Pontificum Amplissima Collectio*, ed. Charles Cocquelines, IV-IV (Rome, 1747), 393.

Among the original duties of the Holy Office was censoring books, until Pius V erected a special congregation for this in 1571, the *Sacra Congregatio Indicis*. The organization of this dicastery that was valid during the interval covered by our report goes back to the bull by Benedict XIV *Sollicita ac provida*, dated July 9, 1753.[367] The bull also contains the order of procedures for the congregation, which, unlike the Holy Office, was structurally no different from the other congregations.

The procedures of the Congregation of the Index were regulated in such a way that its secretary was responsible for deciding whether or not it should deal with a book that had been reported. In order to reach a judgment on this question, first the person who reported it should be interrogated about his reasons, whereupon the secretary had to test the validity of these reasons by reading the book. If the decision was in the affirmative, then he had to appoint two consultors whom the pope or the prefect would assign to prepare expert opinions on the incriminated work. Only if these two agreed did the secretary put the matter on the agenda for the congregation. Now the prefect commissioned a consultor to present a new expert opinion to the "preparatory congregation"—in other words the assembly of consultors that normally met once a month. However, it was not made up of all the consultors, but only of those whom the secretary invited by reason of their special competence.

Now, finally, it was the cardinals' turn; after all, they were responsible for the judgment. They could also take measures that might be suggested by consideration of the author. For example, further expert opinions could be obtained from bishops or faculties,

[367] See SS. D. N. Benedicti Papae XIV, *Bullarium* IV (Rome, 1757), 115–124. See also Joseph Hilgers, *Der Index der verbotenen Bücher: In seiner neuen Fassung dargelegt und rechtlich-historisch gewürdigt* (Freiburg im Breisgau: Herder, 1904), 59–69.

but in particular the author himself or his representative could be interrogated, whereby the latter could even be appointed a sort of public defender by the dicastery itself. The last-mentioned possibilities were to be exhausted, especially when dealing with a meritorious Catholic author in whose work one or another passage was regarded as offensive. Objections could be raised to things in this way, and then the book could be condemned "*donec corrigatur*"[368] (until it was corrected); this decree was not made public, however, and the author had an opportunity to make his views clear. Only if he refused to do this did they proceed to make it public. Moreover, they had to observe the admonition in the bull *Sollicita ac provida* to the effect that the congregation must be strictly nonpartisan in selecting the persons whose writings it dealt with.

It is obvious that in the present case this procedure could be applied only analogously, since it dealt with the opposite of a prohibition: an imprimatur was to be issued against the opinion of the Master of the Sacred Palace, and prohibitions pronounced earlier had to be lifted. In doing so, as will be shown, the two dicasteries worked together closely to reach their common goal. Such a way of proceeding, however, was possible only if the cardinals, officials, and consultors involved were of the same opinion and acted in concert. Therefore, we should take a look at the most important persons who participated.[369]

368 This had happened with Copernicus and his allies in 1616 or 1620. See above, GT96 = around footnote 137. An interesting example of a corrected copy of Didacus a Stunica is found in the Biblioteca Casanatense under the call number KK VII 32.

369 The personnel of the congregations are listed in *Notizie per l'anno 1820 dedicate all'Eminentissimo e Reverendissimo Principe il Signor Cardinale Pietro Vidoni, Diacono di S. Nicola in Carcere* (Rome, 1820). Members of the Inquisition, or the Holy Office: 55–56; members of the Congregation of the Index: 61–63.

Now it is immediately evident that the cardinal-members of the two congregations were for the most part identical. Thus della Somaglia, di Pietro, Pacca, Caselli, della Genga, de Gregorio, Opizzoni, and Fontana were active in both congregations. It should be noted that, in the person of della Somaglia, the secretary of the Holy Office attended sessions of the Congregation of the Index, while the prefect of the latter, di Pietro, was a member of the Holy Office. Galeffi, Brancadoro, Gabrielli, and Consalvi had posts in the Holy Office only, while Cardinals Spina and Ruffo Scilla belonged only to the Congregation of the Index. Of the bishop-consultors, moreover, Bertazzoli and Caprano belonged to both congregations, and of the others, Toni, Demont, and Ostini. Frattini attended sessions of the Holy Office only.

Under these circumstances, it is not surprising that the closest communication imaginable between the two dicasteries was guaranteed, and the scarcity of written records on file is easily explained by the collaboration of Index and Office.

Cardinals and consultors

In the attempt which we now undertake to sketch the personnel of the two congregations, we run very quickly into narrow restrictions — and this should be emphasized from the outset. Besides the excellent *Dizionario biografico degli Italiani,* there is scarcely any biographical reference work for the years in question. Moroni's great work, which offers material on many of the personages who are to be introduced here, is limited in almost all cases to dates of their external career, and leaves us in the dark about their intellectual and Church-political orientations. The *Enciclopedia Cattolica* and the *Enciclopedia Italiana* represent only the state of research in the years in which they were published. A wealth of information and evaluations, in contrast, is the *Tableau des Cardinaux* that the Austrian ambassador in Rome,

Count Apponyi, compiled in 1821 with a view to an imminent conclave and presented to Prince Metternich.[370] Let the attempt be made, therefore, with these resources.

We encounter first the secretary of the Holy Office and a member of the Congregation of the Index, Cardinal Giulio Maria della Somaglia (1744–1830), who held this office from 1814 until his death.[371] A descendant of old Northern Italian nobility, he received the red hat in 1795 as the last of the cardinals created by Pius VI. He demonstrated determination and prudence in numerous offices, some of them in the civil government. He showed fidelity to principle and firmness of character as one of the "black cardinals." He was an influential collaborator in rebuilding the Papal States, and in 1823 Leo XII appointed him Consalvi's successor in the post of cardinal secretary of state. Over the course of the incident that we are about to relate, he also met with Settele, who in his diary calls him not only amiable but also well instructed in astronomy.[372] With firmness and conviction he sided with Settele. He was praised as a perfect man of the world, who possessed intellect and talent; as dean of the Sacred College he kept open house; he had considerable influence on the cardinals, and he was well liked by the high Roman nobility with whom he socialized often — even with Napoleon's family.[373] The story went that in his youth he had been a freethinker, indeed, even one of the Illuminati. He stood in stark

[370] See Haus-, Hof- und Staatsarchiv Wien, Staatskanzlei Rom, Varia 32 (3 vols.). Count Anton Apponyi (1782–1852), friend and patron of the arts and sciences, ambassador to London and Rome, after 1826 in Paris. See Constant von Wurzbach, *Biografisches Lexikon des Kaisertums Oesterreich*, 60 vols (Vienna, 1856–1891), vol. 1 (1856), 57.

[371] See Ritzler and Sefrin, *Hierarchia catholica*, vol. 6, 38 or 87 (on his career until his cardinalate); Moroni, *Dizionario*, vol. 67 (1854), 175–181.

[372] See Settele, August 22, 1820, in Maffei, *Giuseppe Settele*, 342.

[373] See HHStA, Tableau I, fol. 1 v-2 r.

contrast with Consalvi. In the conclave after the death of Pius VII, France and Austria considered him their candidate, and he was able to gather twelve votes in his favor.[374] As secretary of state he proved to be cautious and critical about accusations from Germany that were made against German bishops.[375]

Cardinal Michele Di Pietro (1747–1821) also attended sessions of both congregations.[376] After he had served as professor for theology, canon law, and Church history at the Collegio Romano, Pius VI appointed him titular archbishop of Isauria. During the imprisonment of Pius VI, he was legate and governor of the city of Rome. In 1801 he was created a cardinal *in pectore* (secretly) by Pius VII; the appointment was made public in 1802, and he shared for the second time in the imprisonment of a pope by Napoleon. After Pius VII had signed the Fontainebleau Agreements, di Pietro was one of the pope's counselors who prompted him to revoke his signature. His collaboration in the condemnation of the Synod of Pistoia by the bull *Auctorem fidei,* in the bull of excommunication against Napoleon, and in making arrangements for the conclave before the death of Pius VI show that he was not only a renowned canonist but also a fearless champion of the papal point of view. Di Pietro was close friends with the Barnabite cardinals Gerdil and Fontana and was highly esteemed for his beneficence. He died on July 2, 1821, in Rome.

[374] See Schmidlin, *Papstgeschichte*, index.

[375] See Hermann H. Schwedt, *Das römische Urteil über Georg Hermes (1775–1831), ein Beitrag zur Geschichte der Inquisition im 19. Jahrhundert,* Römische Quartalschrift, Supplementheft 37 (Rome-Freiburg-Vienna, 1980), 34–35.

[376] See Ritzler and Sefrin, *Hierarchia* catholica, vol. 6, 245; vol. 7 (Patavii, 1968), 8–9; Moroni, *Dizionario*, vol. 53 (1851), 37–39; *La Pontificia Università Lateranense, Profilo della sua storia, dei suoi maestri e dei suoi discepoli* (Rome, 1963), 132.

Francesco Luigi Fontana (1750–1822), like two of his brothers, belonged to the Barnabite order; the young cleric was highly gifted, especially literarily, and very successful as a religious teacher and educator.[377] During the French occupation, Fontana, as provincial of his order in Northern Italy, showed extraordinary skill in the most difficult circumstances; this predestined him for the office of general superior of the order, which he assumed in 1807. In 1809 he shared the imprisonment of Pius VII, whom he had already accompanied to Paris for Napoleon's coronation. As a result of his collaboration in the bull of excommunication against Napoleon, he was incarcerated for three years in Vincennes, in the same jail as Opizzoni, di Pietro, and de Gregorio (who will likewise be introduced in this context). After the pope's return, Fontana was created a cardinal in 1816 and became prefect, first of the Congregation of the Index, then of the Propaganda Fide. Beyond the circle of his confreres, Cardinals Gerdil and Lambruschini, Fontana was close friends with Capellari and Manzoni. His scholarly interests prompted him to collaborate also in the Accademia di Religione Cattolica, of which he was the secretary. At this stage, however, Fontana, who had been considered the most learned among the cardinals and always as *papabile* (a good candidate for pope), was sick and had weak nerves, so that he could scarcely take an interest in business any more.[378]

We just mentioned Cardinal Emmanuele de Gregorio (1758–1839), who was a descendant of the high-ranking nobility of Regno;

[377] See Ritzler and Sefrin, *Hierarchia catholica*, vol. 7, 12; Moroni, *Dizionario*, vol. 25 (1844), 150–155; Piolanti, *L'Accademia di Religione Cattolica*, 51 and index; Giuseppe Boffito, *Scrittori Barnabiti o della Congregazione dei Chierici Regolari di San Paolo (1533–1933)*, Biografia, Bibliografia, Iconografia, vol. 2 (Florence: Olschki, 1935), 35–50.

[378] See HHStA, Tableau II, fol. 10 v-11 r.

his mother came from Barcelona.[379] While his brothers, who were raised with him in Rome, embarked on a military career, Emmanuele became a priest and entered the service of Pius VI, whose fate he shared during the Napoleonic era. During the imprisonment of Pius VII and his deputy di Pietro, de Gregorio conducted the business of the supreme administration of the Church as well as was possible under those circumstances, until he too, was summoned to Paris and repeatedly incarcerated. From 1816 on, as a cardinal, he held several administrative offices in the Curia. He was renowned as a comprehensively trained theologian and jurist, and was praised for his straightforward integrity.[380] Consalvi, on the other hand, called him — according to Count Apponyi — uneducated and ambitious; this, however, was certainly the expression of a dislike, which incidentally was mutual. The twenty-four votes that he was able to gather in the conclave after the death of Leo XII were a sign of his great prestige.

It seems that in the present context we do not need a separate appreciation of Cardinal Carlo Francesco Caselli (1740–1828), who did play a significant role in connection with the concordat with Napoleon and the subsequent events, although it was not always an unambiguous one; after 1814, however, he no longer took an interest in the matter.[381] Our Austrian informant certifies that he had the highest qualities of intellect and character, as well as the unlimited respect of all, especially of the other cardinals.[382] In Apponyi's

[379] See Ritzler and Sefrin, *Hierarchia catholica*, vol. 7, 11 and index; Moroni, *Dizionario*, vol. 33 (1845), 10–16; M. Caffiero, in *DBI* 36 (1988): 212–215; R. Aubert, in *DHGE* 22 (1988): 88–91

[380] See HHStA, Tableau II, fol. 9 r-10 v.

[381] See Stefano Da Campagnola, in *DBI* 21 (1978): 320–323; Pietro Benassi, "La formazione culturale del card. C. F. Caselli (1749–1828)," *Studi Storici dell'Ordine dei Servi di Maria* 30 (1980): 139–203.

[382] See HHStA, Tableau II, fol. 4 rv.

opinion, Caselli would be an excellent pope, if his advanced age and his close ties with France had not stood in the way.

Because they are generally well known, Ercole Consalvi, Bartolomeo Pacca, and Annibale della Genga (the future Leo XII) need no introduction. In contrast, the following names are less familiar.

Pietro Francesco Galeffi (1770–1837), already a cardinal at the age of thirty-three, was one of the "black cardinals" during the Parisian imprisonment; in the year 1820 he was archpriest of St. Peter's Basilica and cardinal bishop of Albano.[383] He had the reputation of being a zealous shepherd of souls. He was also praised for his love of neighbor. This is in keeping with Count Apponyi's description of him as an irreproachable, very well-mannered priest.[384] However, the Austrian ambassador thought it necessary to certify also his slight education and limited horizon. As an extreme partisan of the "*zelanti*" (zealots), Galeffi was reportedly also an opponent of Consalvi. He is said to have owed his career to his ties of kinship with Pius VI.

Giulio Gabrielli (1748–1822), created a cardinal by Pius VII in 1800 in his first promotion, was pro-datary in the year 1820.[385] Under Napoleon he too had to endure imprisonment repeatedly, especially since he likewise was one of the "black cardinals." After his return to Rome, Gabrielli was one of the first members of the newly instituted Congregation for Extraordinary Ecclesiastical Affairs, together with della Somaglia, di Pietro, Pacca, Brancadoro, and three other cardinals who were not concerned with the matter we are discussing here. To the ambassador from the Austrian court, Gabrielli seemed witty and talented, but also of strong and

[383] See M. DeCamillis, in *EC* 5 (1950): 1885; Ritzler and Sefrin, *Hierarchia catholica*, vol. 7, 10.

[384] See HHStA, Tableau I, fol. 6 r -7 v.

[385] See Schmidlin, *Papstgeschichte*, index; Ritzler and Sefrin, *Hierarchia catholica*, vol. 7, 7.

dogmatic character.[386] Although indebted to Metternich's political ideas, he criticized the Austrian state Church quite openly. Apponyi estimates that his influence will be significant in a future conclave, too, whereby the *zelanti* would count on him. However, the fact that one of his nephews married a daughter of Lucien Bonaparte — along with his difficult, avaricious character — lessened his chances of being elected.

Cesare Brancadoro (1755–1837) also belonged to the Holy Office.[387] In 1789 he was appointed titular archbishop of Nisibis; subsequently, while serving on a diplomatic mission in Belgium and Holland, he tried in vain to ally the clergy and people of Belgium with the Austrian government against the pressing forces of the Revolution. In 1801 he became a cardinal and bishop, first of Orvieto, then archbishop of Fermo, his hometown. When Pius VII was deported to Paris, he too was among the "black cardinals," and in the aftermath he was in favor of a rigorous restoration and one of the leading *zelanti*, unlike Consalvi. It is not surprising, then, to find him among the staunch opponents of those who advocated renouncing the Church properties that had been seized over the course of the revolutionary incidents.[388] Both his education and his literary and rhetorical abilities were generally praised; he himself composed several apologetic and belletristic works.

Another member of the Congregation of the Index was Cardinal Giuseppe Spina (1756–1828).[389] A close collaborator with Pius VI who accompanied the pope into French exile, he became

386 See HHStA, Tableau II, fol. 3 r-4 r.

387 See G. Pignatelli, in *DBI* 13 (1971): 801–804.

388 See HHStA, Tableau II, fol. 2 rv.

389 See Ritzler and Sefrin, *Hierarchia catholica*, vol. 7, 8; S. Furlani, in *EC* 11 (1953): 1123; Moroni, *Dizionario*, vol. 68 (1854), 280–285; Schmidlin, *Papstgeschichte*, index.

historically important when alongside Consalvi he prepared the concordat with Napoleon and cosigned it in 1801. In the same year he was named a cardinal *in pectore*; the appointment was made public in 1802 and he became archbishop of Genoa. While with Pius VII in Paris, he sided with the "red cardinals" in 1810 and even participated in the National Council commanded by Napoleon. He remained a supporter of Napoleon, which did not prevent him from receiving high offices in Rome after 1816. In 1820 he was cardinal bishop of Palestrina and legate of Bologna. The reason for this was certainly that generally he was highly esteemed for his qualities of intellect and character, which he demonstrated as legate in Forli and especially in Bologna.[390] He had a long and close friendship with Consalvi.

Cardinal Luigi Ruffo Scilla (1750–1832) was descended from a princely family in the Regno province.[391] At first titular archbishop of Apamea, in 1801 he became a cardinal and in the following year archbishop of Naples. Driven from there by the French, he had to go with Pius VII into exile to France, where he was interned in undignified circumstances in Saint-Quentin; he too was a "black cardinal." Our Austrian informant very critically calls him deficient in mind and heart, and also an ignorant fanatic.[392] We learn from him also that Ruffo Scilla took the oath to the revolutionary constitution of Naples. His prestige, both within and outside the Sacred College, was so low that he had no chance whatsoever in a conclave. After the Restoration he was again archbishop of Naples, and he probably did not participate actively in the works of the Congregation of the Index.

[390] See HHStA, Tableau I, fol. 4 r-5 v.

[391] See Ritzler and Sefrin, *Hierarchia catholica*, vol. 7, 7; Moroni, *Dizionario*, vol. 59 (1852): 220–222.

[392] See HHStA, Tableau II, fol. 1 rv.

Cardinal Carlo Opizzoni (1769–1855) may have been in a similar situation.[393] In his youth a freethinker and a libertine surrounded by scandals, at the age of thirty-three he was appointed archbishop of Bologna by Napoleon.[394] He also owed his cardinalate to the Corsican. Moreover, the Viennese diplomat depicts him as an opportunist with no character. Yet the fact that during the exile Opizzoni was one of the "black cardinals" warns us to be cautious about such judgments. Still, even Count Apponyi reckoned him as part of the moderate wing of the *zelanti*. Only after 1831 did the highly regarded cardinal become politically active. In 1849 the sick archbishop succeeded in obtaining an honorable capitulation of Bologna to the Austrians.

More important than the prelates just mentioned, however, were the consultors. Among them, as members of both Congregations, were the titular bishops Bertazzoli and Caprano.

Francesco Bertazzoli (1754–1830) was from Lugo.[395] He was ordained a priest at the age of twenty and, thanks to his excellent education, very soon taught as a professor at the seminary in Lugo. He also joined the Arcadia and several other academies in Bologna. In 1796 he helped the future Pius VII, then bishop of Lugo, to end an anti-French revolt. In 1799 Bertazzoli financed Chiaramonti, who was destitute on account of his support for French emigrant priests, so that he could take part in the conclave in Venice, from which he emerged as Pius VII. After that, as titular archbishop of Edessa and *elemosiniere* (manager of papal charities), Bertazzoli was

393 See Ritzler and Sefrin, *Hierarchia catholica*, vol. 7, 10, 114–115; M. De Camillis, in: *EC* 9 (1952): 170.

394 See HHStA, Tableau II, fol. 4 v-5 r.

395 See R. Colapietra, in *DBI* 9 (1967): 483–484. See also Piolanti, *L'Accademia di Religione Cattolica*, index, regarding his activity for the Accademia di Religione Cattolica, esp. 33–34; Bidolli, "Contributo," 82–83.

constantly at the pope's service; he also accompanied him to Paris for Napoleon's coronation. His promotion to cardinal may have been prevented by both Pacca and Consalvi, who considered him rather naive despite his outstanding education. He showed weakness of character in dealing with Napoleon, who by threats and arrest induced him to get Pius VII to put his signature under the shameful "Concordat" of Fontainebleau. After the Restoration, Bertazzoli was president of the Accademia di Religione Cattolica. Together with Cardinal Pacca, he deserves considerable credit for the success of the reform of studies and universities in the Restoration years. Finally made a cardinal in 1823, during the conclave of that same year he was among the promoters of Genga, who then appointed him prefect of the Congregation for Studies. In this capacity he struck up a close friendship with the future Gregory XVI. He belonged to the group of *zelanti* and was an opponent of Consalvi's reforms.

Pietro Caprano (1759–1834), also a Roman, played a not-unimportant role in the development we will describe.[396] Against his family's will, after his studies with the Jesuits he became a priest in 1782, having earned a doctorate in theology in 1780. At the Collegio Romano he taught liturgy, moral theology, and Church history successively. Because he, unlike Settele, refused to take the oath to the regime of the French occupation, he was deposed and, from 1809 to 1814, deported to Northern Italy as being politically suspect. After the return of the pope, he became his private chamberlain and assumed several curial offices. In 1816 he was appointed titular archbishop of Iconium, and in 1822 secretary of the important Congregation for Extraordinary Ecclesiastical Affairs, an office that he lost when Consalvi was removed from the Secretariat of State. In

[396] See F. Raco, in *DBI* 19 (1976): 163–165.

1820 Caprano was assigned to the delicate task of elaborating Rome's position on de Maistre's famous work *Du pape* (On the pope), which had appeared the previous year in Lyons. In 1823 he became secretary of the Propaganda Fide and thus collaborated again with Consalvi. In 1826 he was made a cardinal *in pectore*; the appointment was made public in 1828. In the 1831 conclave he was an opponent of the election of Capellari.

The consultors Toni and Demont do not appear in connection with the Settele affair; there will be a report on Ostini elsewhere.[397]

If one surveys the available information about the curricula vitae, education, and professional activity or pastoral and political experience of the personages involved in the decisions by the Holy See that are of interest here, what meets the eye first is the fact that they all lived through the Napoleonic troubles and suffered because of them. In these events, which radically changed the circumstances of their lives — loss of freedom, the most undignified treatment, and bitter poverty, which made even obtaining daily sustenance dependent on charitable gifts — they all, with the sole exceptions of Spina and Caselli, demonstrated a loyalty to the pope and to the Church that bordered on the heroic. All this is an excellent testimonial to their character.

With regard to their education, most of them prove to be generalists, who had had a thorough, primarily philological-literary, and of course also philosophical-theological, juridical, and historical universal education, which made them capable of acquiring in-depth knowledge from one case to the next. Interest in the natural sciences seems to have been cultivated the least; Settele's praise of the astronomical knowledge of Cardinal della Somaglia seems to be an exceptional case.

[397] See GT185 in the subsection "New Obstacles," around footnotes 64-66.

However, since the problem to be discussed here was not primarily an astronomical one, all who dealt with it were no doubt equal to their task.[398]

[398] See Christoph Weber, *Kardinäle und Prälaten in den letzten Jahrzehnten des Kirchenstaates: Elite-Rekrutierung, Karrieremuster und soziale Zusammensetzung der kurialen Führungsschicht zur Zeit Pius' IX. (1846–1878)* I, Päpste und Papsttum 13,1 (Stuttgart: Hiersemann, 1978), 147–151.

CHAPTER 3
DRAMATIS PERSONAE

GIUSEPPE SETTELE

CREDIT FOR HAVING initiated the definitive resolution of the Galileo case goes to the professor of optics and astronomy at the Sapienza and canon of Santi Celso e Giuliano in Banchi,[399] Giuseppe Settele.[400]

Settele was born in Rome on December 30, 1770, son of the baker Xaver Settele. The father was originally from Seeg, located near Allgovia (Swabia, Southern Germany), where he had been born in 1730, and he followed several family members—almost all of them bakers—who over the course of the eighteenth century had moved to Rome. There he owned a bakery in the Trastevere district

399 See Walther Buchowiecki, *Handbuch der Kirchen Roms,* vol. 1 (Vienna: Hollinek, 1967), 519–525.

400 For a short biography, see Maffei, *Giuseppe Settele,* 25–40. The main source for his biography is Settele's diary (see below, notes 9 f.), which Vernacchia-Galli drew on for the period from 1810 to 1841 with a view to a history of the Sapienza. Unfortunately that study lacks a scholarly apparatus citing secondary literature. Maffei reprints the passages from the diary that refer to the present topic "Copernicus and Galileo." *Giuseppe Settele,* 285–421. His reprint likewise suffers from the fact that he merely offers a text without commentary, one which moreover contains numerous erroneous readings. Unless otherwise cited, we follow the two works just mentioned in our short biographical sketch.

when in 1767 he married the baker's daughter Therese Hipp, who was eighteen years younger.[401]

Their son Giuseppe first learned his father's trade in his shop, but then chose the clerical state, less for religious reasons than for the sake of the opportunity for higher education; as a student of the renowned Gioacchino Pessuti[402] and soon of Feliciano Scarpellini,[403] too, he began at a very early age to take an interest in astronomy, physics, and mathematics, but also in the science of archeology that was developing at the time. In 1801 he was one of the re-founders of the Lincean Academy, and he soon joined the Accademia dell'Arcadia, too.

Settele, too, was immediately caught up in the political turbulence of the Napoleonic era; meanwhile (exactly when is not known) he had become a professor at the Sapienza. Now in 1812, when he too was ordered to take the oath of loyalty to the constitution of the French Empire, which Pius VII had explicitly forbidden, Settele was plunged into a major conflict of conscience. Eventually, under the pressure of

[401] About the Settele/Hipp family see *Kölnische Zeitung/Abend-Ausgabe* (Cologne), December 15, 1902, no pagination; Friedrich Noack, *Das Deutschtum in Rom seit dem Ausgang des Mittelalters* II (Stuttgart: Deutsche Verlagsanstalt, 1927; repr. Aalen, 1974), 554; Albrecht Weiland, *Der Campo Santo Teutonico in Rom und seine Grabdenkmäler*, Der Campo Santo Teutonico in Rom, vol. 1 (Rome-Freiburg-Vienna: Herder, 1988), 538–540. The point of departure of the Settele family was the hamlet Settele in eastern Allgovia that was named after it. See G. Reiprich, "Die Familiengeschichte," in Herbert Buchner, Ludwig Feigl, Gert Reiprich, *Settele, Geschichte einer schwäbischen Familie* (Augsburg: Brigitte Settele Verlag, 1989), 11–12.

[402] Gioacchino Pessuti (1743–1814), mathematician. Settele was his successor at the Sapienza. See Giuseppe Ferretto, *Note storico-bibliografiche di Archeologia cristiana* (Vatican City: Poliglotta Vaticana, 1942), 299, 459.

[403] Feliciano Scarpellini (1762–1840), astronomer. See *La Pontificia Università Lateranense*, 284; and G. Armellini, in *Enciclopedia Italiana di scienze, lettere ed arti*, 36 vols. (hereinafter EncIt) (Rome, 1929–1939), vol. 31 (1936), 13.

his financial situation, he agreed to comply with the order, so as not to lose his position. His diary testifies that this decision left him with pangs of conscience for a long time. The same is true about the economic troubles caused by his family situation, so that occasionally he could barely earn a living. His salary covered less than half of his expenses, even when these were restricted to what was absolutely necessary; his financial worries were diminished only when he managed to increase his income by giving private lessons. His most prominent student was the then-thirteen-year-old son of Louis Napoleon. One endearing feature of Settele's character is that, despite his bitter personal indigence, he cared for his relatives in distress and shared with them his meager supply of bread. These same circumstances probably kept him from engaging in more extensive literary activity or having more frequent social contacts, which might have broadened the very narrow circle of his life.[404] Still, Settele maintained good relations with several colleagues, which would later be to his advantage in the controversy about Copernicus and Galileo. Nevertheless, we cannot overlook a certain narrowness of Professor Settele's human, intellectual horizon. No stronger religious impulses prompted him to go beyond the modest standard of conventional piety.

What made him really prominent was, surprisingly, not astronomy or physics, but rather his successes in archeology. In the last third of his life, Settele in fact made several groundbreaking discoveries in that field. In this connection, then, we find his name, too, mentioned in Cardinal Wiseman's *Memoirs*, from which he made his way into

[404] A provisional list of his writings can be found in Maffei, *Giuseppe Settele*, 41–43; for observations about unpublished manuscripts, see Vernacchia-Galli, *L'Archiginnasio Romano*, 17; G. Ferretto, *Note storico-bibliografiche di Archeologia cristiana* (Vatican City: Poliglotta Vaticana, 1942), 299–301. See the earlier source, Giovanni Battista de Rossi, *Roma sotterranea* I (Rome: Cromo-Litografia Pontificia, 1864), 63–64.

Schmidlin's *History of the Popes,* and finally into Jedin's *Manual of Church History.*[405] Settele was highly esteemed as an archeologist by Leo XII and especially by Gregory XVI. The latter already knew Settele from the time of the conflict that will be described here, during the course of which he had issued an opinion in favor of the professor.

Settele became famous above all through his investigation of the Vatican grottos, during which he succeeded in finding the epitaphs of Boniface II, Gregory the Great, Sabinian, and Hadrian II. But those were only his most spectacular discoveries. They are the subject matter of the folio volume that he wrote together with Emilio Sarti: *Ad Philippi Laurentii Dionysii Opus de Vaticanis Cryptis Appendix in qua nova Cryptarum Ichnographica Tabula adiectis notis inlustratur auctoribus Aemiliano Sarti et Josepho Settele in Romano Archigymnasio professoribus* (appendix to the work by Filippo Lorenzo Dionisi "On the Vatican Crypts," in which a new scale map of the crypts is explained with additional notes by authors E. S. and G. S., professors at the Sapienza), published in Rome by Iohannis Ferretti in 1840.

At the same time, this was also Settele's last work. He died on March 6, 1841,[406] and was laid to rest in the tomb of his family on the Campo Santo Teutonico.

Maurizio Benedetto Olivieri

Far more significant than the role of the author Settele was the one that the commissioner of the Holy Office, Fr. Maurizio Benedetto

[405] See *Memoir of His Eminence Cardinal Wiseman* (London: Richardson and Son, 1865). See Schmidlin, *Papstgeschichte,* 174–175. See Roger Aubert, "Licht und Schatten der katholischen Vitalität," in Hubert Jedin, ed., *Handbuch der Kirchengeschichte* VI/1 (Freiburg-Basel-Vienna: Herder, 1971), 674.

[406] On the date of death, see F. Noack, *Das Deutschtum in Rom,* 2:554; Albrecht Weiland, *Der Campo Santo Teutonico in Rom,* 540 (a copy of the tombstone inscription noting the day of death is in ibid., 538).

Olivieri, played in the solution to the problem.[407] Almost exactly the same age as Settele, he was born on February 24, 1769, in Accelle, Savoy, and entered the Dominican order at the age of only fifteen. After teaching philosophy at the schools of his order in his younger years, he studied Hebrew philology and also was occupied with biblical and Old Testament, text-critical questions. Finally he became professor for Old Testament studies at the Sapienza and taught also at the Propaganda College. On May 8, 1806, he was accepted into the Accademia di Religione Cattolica to replace the late Cardinal Corsi; as a member he not only held important offices but also gave a number of lectures.[408] Olivieri seems not to have participated much in the activities of the Accademia dell'Arcadia or the literary academies of Florence and Milan, to which he also belonged.[409] In the years 1834–1835 Olivieri was for a short time the general of his order. He resigned from the office after a year because he did not agree with the ideas of Gregory XVI about the organization of the generalate.[410]

His professional life was actually centered on his academic teaching and his office as commissioner of the Holy Office, which he held for a quarter of a century. Within this framework Olivieri also developed his literary activity, which dealt with biblical-philological, theological, and historical themes. At the age of twenty-four he seems to have earned a doctorate in theology at the University of

407 An obituary for Olivieri penned by the Franciscan Fr. Bernardino da Ferentino is found in the Archivum Generale Ordinis Praedicatorum, Santa Sabina, Rome, V 29. About Olivieri see also Moroni, *Dizionario*, vol. 13 (1842), vol. 132; vol. 55 (1852), 96; vol. 60 (1853), 305–306; Innocenzo Taurisano, *Hierarchia Ordinis Praedicatorum*, vol. 1 (Rome, 1916), 15, 77.

408 See Piolanti, *L'Accademia di Religione Cattolica*, 94 and index.

409 In the *Accademia dell'Arcadia* he went by the name Anassimandro Lisio. A very dry poem by him can be found in the *Giornale Arcadico di Scienze* 6 (1826): 33.

410 See Daniel Antonin Mortier, *Histoire des Maires Généraux de l'ordre des Frères Prêcheurs* VII (Paris: Alphonse Picard et fils, 1914), 475.

Parma with a disputation about the Hebrew text of the Bible.[411] As his next publication, we find a short treatise about the need to cultivate the classical languages,[412] and in the following year Olivieri presented his most voluminous work, of more than four hundred pages: a Church history of the eighteenth century.[413] Probably an acute awareness that he lived in a time of far-reaching upheavals was what compelled his alert mind to investigate the roots of what he was experiencing. At that time Olivieri was exactly thirty-eight years old, and his work was published during the exile of the pope in France.

Olivieri organizes his fact-filled presentation chronologically by pontificates, whereby he assigns particular value to intellectual developments. It is interesting that he views Montesquieu and Rousseau as forerunners of the French Revolution and also devotes critical discourses to the Illuminati. He has remarkable sympathy for the Jesuit order, which had been suppressed in 1773. Thus, already as a relatively young man, the author of this "history of an ecclesiastical era" proves to be an observer whose judgments are quite independent. One result of his exegetical-philological work was a study refuting the remarks of the prefect of the Vatican Library, Francesco Antonio Baldi, who claimed to have found in the Hebrew text of the Bible hitherto unnoticed prophetic references to the Cross of Christ.[414]

411 See his work "De sacro hebraico textu . . . disputationem publice instituit fr. M. B. Olivieri," (Parma, 1793), which runs to 77 pages (Biblioteca Apostolica Vaticana, BAV, Ferraioli, III, 1729).

412 "De linguarum eruditarum cultu graviorum disciplinarum studiis jungendo" (Rome, 1806), 16 pages (BAV Mai X-F-VI 4 int. 48).

413 *Storia ecclesiastica del secolo decimo ottavo* (Rome, 1808), is 408 pages long.

414 Francesco Antonio Baldi (ca. 1750–1826), recording secretary of the Secretariat of State; 1802, member of the Accademia di Religione Cattolica; 1814–1818, prefect of the Vatican. His book *De apologia catholicae religionis* (Venice, 1799) had already met with caustic criticism. The same happened with the short work *Incognitorum hactenus vaticiniorum*

An example of "recreational Hebrew" was an article published in the *Notizie del Giorno,* in which Olivieri had attempted to translate verses from Dante into Hebrew and Greek, and which was associated with a minor controversy.[415]

Of a more serious nature, however, was another controversy that occupied philosophers and theologians in Rome at that time. It concerned the abbot and professor of philosophy Marco Mastrofini and his work *Metaphysica sublimior de Deo Trino et Uno,* which was published in 1816 in Rome and made the claim that it could prove the Divine Trinity philosophically, whereby Mastrofini came very close to semi-rationalism. Olivieri, too, grappled very critically with the book, and finally rejected Mastrofini's position.[416] This literary

de Cruce ... interpretatio ex Hebraeo et declaratio (Rome, 1817). Olivieri opposed his arguments in a fifty-four-page treatise *De voce — Chen in truncum et trunco in crucem versis, unde incognita hactenus cruce vaticinia in sacro hebraico textu clarissimus vir Franciscus Antonius Baldi ... a se detecta exhibuit lucubratiuncula Fr. M. B. Olivieri* (Rome, 1817). Because of this controversy Olivieri had already come into conflict with Anfossi. See also his own statements to Anfossi in his letter dated May 24, 1821, in Brandmüller and Greipl, *Copernico, Galilei e la Chiesa,* 397–405.

415 See *Notizie del Giorno,* June 11, 1819, 4. D. Ricci replied to this with the nineteen-page essay "Al molto reverendo Padre M. Olivieri professore di Ebraico nell'archiginnasio Romano sull'articolo letterario pubblicato nelle Notizie del Giorno nr. 21," (Rome, 1819). On the BAV copy there is a handwritten note by the poet Giuseppe Gioacchino Belli (1791–1863): "This reply was given to me on May 4, 1835 by Prof. Lanci together with his preceding dissertation; very few copies of this reply are in circulation, since Lanci did not want Fr. Olivieri to be annoyed." About Belli, see G. Orioli, in *DBI* 7 (1965): 660–668.

416 The full title is *Metaphysica sublimior de Deo Trino et Uno, auctore M. Mastrofini presbytero, matheseos olim ac philosophiae Publico Professore I* (Rome, 1816). Olivieri's 122-page treatise had the title *Dell'opera intitolata 'Metaphysica sublimior de Deo Trino et Uno, auctore Marco Mastrofini presbytero, matheseos olim ac philosophiae Publico Professore' Saggio critico* (Rome, 1821). About Marco Mastrofini (1763–1845), see Carlo Gazola, *Alla onorata memoria dell'Abate D. Marco Mastrofini prosa* (Rome: Tipografia

activity went hand in hand with Olivieri's involvement in the program of the Accademia di Religione Cattolica. It had set for itself the task of countering the challenges by Enlightenment rationalism with an apologetic that would be adequate for the times. Within the framework of this program, Olivieri as an Old Testament scholar naturally dedicated himself to the problems that resulted from the confrontation of biblical prehistory with the findings of the contemporary sciences. His method in this undertaking consisted above all in demonstrating that the two are compatible. However, he also treated exegetical topics in the strict sense.[417]

---

delle Belle arti, 1845); Hugo Hurter, *Nomenclator Literarius Theologiae Catholicae: Theologos exhibens aetate, natione, disciplinis distinctos*, V/1 (Oeniponte: Libraria Academica Wagneriana, 1911), 1188; Saturnino Ciuffa, *Marco Mastrofini, sue opere edite ed inedite e suoi contradittori: Memorie storico-apologetiche* (Rome, 1875); and F. dell'Addolorata, in *EC* 8 (1952): 328.

417 Topics of Olivieri's lectures to the Accademia di Religione Cattolica include the following (translated from Italian): "Bonnet's palingenesis, and other conjectures and systems of the modern naturalists, that rely on the metaphor of the grain of wheat adapted from St. Paul 1 Cor 15:37, are not capable of explaining naturally the resurrection of the body." "Geology can offer assistance to understand how what is related in Sacred Scripture regarding the earthly Paradise and other antediluvian localities could have existed." "Whether Religion has preserved antediluvian writings for us." "The anomalies provided by measuring various [astronomical] meridian lines can serve as an argument for some catastrophe through which the earth's axis had changed its situation; and can give a probable explanation of the deposits belonging to tropical climes which are found in glacial climes: which great catastrophe had characteristics that coincide with those recorded by Moses in the history of the Universal Deluge." "The most exact, latest physical-climactic theories about the atmosphere, and about the composition and decomposition of water provide us with a way of explaining: a) how some preexisting elements could, at the Divine command, increase the water until it surpassed the highest mountaintops over the whole surface of the earth in the Universal Deluge; b) how they diminished without being annihilated until they reached their present postdiluvian level." Piolanti, *L'Accademia di Religione Cattolica*, 157, 163, 514, 172,173.

Finally, in one of his lectures in 1841 he dealt with the relation between Copernicus and the Church. The topic was "Contributions of the Roman Pontiffs to Astronomy."[418] The manuscript was expanded extensively and, long after Olivieri's death on September 27, 1845, was published by his confrere Tommaso Bonora in 1872 in Bologna.[419]

As a public figure, Olivieri was highly esteemed, indeed revered by his contemporaries. Moroni, who was intimately familiar with the Roman scene during those years, calls him one of the most learned men of his time, which is certainly exaggerated, yet does indicate something of the reputation that he enjoyed. Particularly valuable, though, is Moroni's remark that Olivieri's erudition was combined with the abundant gifts of a big heart.[420] This judgment is corroborated by a characterization that comes from an altogether unexpected quarter. Of all people, the former qualificator of the Holy Office, Luigi Desanctis, who in 1847 had become a Protestant, found friendly words to write about Olivieri in his unbelievable lampoon *Roma papale*. He writes that the prisons of the Inquisition were filled once again shortly after Pius IX started his pontificate, after they had been almost empty during the reign of Gregory XVI. The previous leniency had been due to Olivieri, "a gifted and very liberal man, who did everything he could to mitigate the rigor of that terrible

[418] See Piolanti, *L'Accademia di Religione Cattolica*, 185.

[419] See M. B. Olivieri, *Di Copernico e di Galileo*. The publisher's name is noted in handwriting in the catalogue of the Biblioteca Casanatense. The remark in the Vatican Library's catalogue that the work was not distributed after it was printed is inaccurate. After all, in that same year it was possible to publish a response to it: Gilberto Govi, *Il S. Offizio, Copernico e Galileo, a proposito di un opusculo postumo del P. Olivieri sullo stesso argomento* (Turin: Appunti, 1872).

[420] See Moroni, *Dizionario*, vol. 60 (1853), 305–306.

tribunal."[421] Thus Olivieri's intellectual and personal qualities are equally attested by such contrasting witnesses. Their judgment will be more than confirmed by our findings.

Fabrizio Turiozzi

A close collaborator with Olivieri was the assessor of the Holy Office, Msgr. Fabrizio Turiozzi.[422] Son of a noble family resident in Toscanella, Viterbo, Fabrizio was born on November 16, 1755. He completed his studies at the seminary in Montefiascone and then in Rome, where he studied jurisprudence and theology. Next he was made a prelate for employment as a diplomat. In 1797 he represented the Holy See at the peace negotiations in Rastatt. In 1802 Pius VII appointed him referendar of the *Segnatura di Grazia papale e della Segnatura di Giustizia* (the predecessor of today's Supreme Tribunal of the Aposotilic Signatura) and finally governor of Jesi, from which he was banished by the French occupier, General Miollis, to his native Toscanella. After the end of Napoleonic rule, having returned to his post, he became — now an apostolic prothonotary — a delegate of Frosinone, and in 1816, succeeding Malvasia who had been appointed a cardinal, assessor of the Holy Office and a canon of St. Peter's Basilica. We encounter Turiozzi in this capacity in the present context. As early as 1823 he also became cardinal of Santa Maria in Aracoeli. In the conclave after the death of Pius VII he was able to garner nine and then thirteen of the fifty-seven votes.[423] For health

[421] "uomo dotto e molto liberale, che per quanto era in lui addolciva il rigore di quel terribile tribunale." Luigi Desanctis, *Roma papale* (Florence: Tipografia Claudiana, 1871), 121. About Luigi Desanctis (1808–1869), see Schwedt, *Das römische Urteil*, 133; C. Crivelli, in *EC* 4 (1950): 1462–143.

[422] For lack of a modern, critical presentation, we rely on Moroni, *Dizionario*, vol. 81 (1856), 476–478. The dates of his hierarchical career are provided in Ritzler and Sefrin, *Hierarchia catholica*, vol. 7, 15.

[423] Albani and Consalvi were the ones who promoted his candidacy. See Schmidlin, *Papstgeschichte*, 372.

reasons he was unable to accept an appointment as legate of Bologna. He died at the age of almost seventy-one on November 9, 1826.

In Turiozzi we find a typical representative of the curial prelature of those years. Educated, but not a scholar, talented, prudent, and experienced, he intervened benevolently in the development of the case of Settele, who mentions him several times in his diary with appreciation and expressions of high esteem on account of his personal qualities. He describes him as entirely without prejudice, well educated, talented, and extremely well-mannered. He also complains about the imprudence and unworldliness of the friars.[424] Although Turiozzi was interested in scholarship, he himself published nothing, as far as we know.

Alessandro Angelico Bardani

Siding completely with Olivieri and Turiozzi was the secretary of the Congregation of the Index, Fr. Alessandro Angelico Bardani, a Dominican, like Olivieri.[425]

Born around 1761 in Castello Azzara, Siena, he entered the Dominican order and soon held high offices, in which he was able to win the high esteem of the Tuscan bishops. Finally his career path led him to Rome, where for two years he was socius (associate) of the Master of the Sacred Palace, until in 1819 he became secretary of the Congregation of the Index; he remained in that post until his death on June 11, 1832. Within his order he was appointed pro-vicar general in 1823. Among the better-known literary works that he penned are a paraphrase of the Psalms, the conversion story of a Calvinist named John Hyacinth Kerfbyl, and a work about the Synod of

[424] See Settele, January 21, 1820, in Maffei, *Giuseppe Settele*, 292.

[425] See Taurisano, *Hierarchia Ordinis Praedicatorum*, 120; Schwedt, *Das römische Urteil*, 23.

Pistoia.[426] The most significant thing about him, however, is an extremely unusual fact. Namely, Bardani recorded in an official file of the congregation his judgment about the proceedings in the Settele affair and thus about his confrere Anfossi and — this is the really interesting part — the role of Pius VII in this matter. And it is by no means a flattering judgment.[427]

About Bardani himself we have a characterization in the form of an obituary, which was composed by a theology professor from Bonn, Johann Wilhelm Josef Braun:

> Lord Alexander Bardani, Secretary of the Congregation of the Index, passed away on June 11 at the age of 71. For twelve years he administered conscientiously at this post as he thought best. Bardani combined simplicity with shrewdness, as the Gospel recommends; he was pious and irreproachable in his morals and in his life. He joined extensive theological knowledge with an utterly unassuming character; although living in cloistered seclusion, he followed the life and the paths and arts of the world; he was familiar with the state and the trends of theology in the Catholic countries, and even German theology remained less foreign to him than to most Romans already working in higher offices; the perfidious tricks that are so frequently attempted from those heights often ran

[426] *Psalterium davidicum syntactica paraphrasi juxta textum* (Rome, 1830); "Lettera ad un amico nella quale si espone la vera conversione alla fede cattolica dal calvinismo e la preziosa morte del Sig. Giovanni Giacinto Kerfbyl..." (Livorno/Leghorn, 1807), 120 pages (BAV R. G. Theod. IV, 3810); *Osservazioni sopra l'articolo intitolato: Détails historiques sur la condamnation du Synode de Pistoie* (Paris, 1821), 191 pages.

[427] Brandmüller and Greipl, *Copernico, Galilei e la Chiesa*, 482–483.

> aground on his discreet deliberation and experienced prudence; not infrequently he attended to those who had been charged with heresy, corrected their false views sparingly and lovingly, and defended the accused themselves against the often strong and stubborn desire to declare them heretics. He carried out several others missions faithfully and to the satisfaction of his superiors; he left behind various valuable writings on ecclesial subjects. His death is a loss for the Institute which he served preeminently and sincerely hoped to reform in a timely way.[428]

Filippo Anfossi

The fact that Mario Rosa devoted a short biography to him[429] makes it easy for us to introduce now also the man who actually caused the whole Settele-Copernicus-Galileo affair and complicated it exceedingly: Fr. Filippo Anfossi, O.P., *Magister Sacri Palatii*. Born in Taggia, Imperia, in 1748, Anfossi entered the Dominican order at an early age; he worked as a teacher at its schools and moreover as a preacher. His scholarly-literary activity developed only at a mature age, during the years after the French Revolution. Recurring topics in it are (1) debating the evolutionism, materialism, and progressive optimism of the Swiss philosopher Charles Bonnet, (2) propaganda against the ideas of the Revolution, and especially (3) Jansenism and the

[428] *Zeitschrift für Philosophie und katholische Theologie* 3 (1832): 209–210. About Johann Wilhelm Josef Braun (1801–1863), see J. R. Geiselmann, in *LThK2* 2 (1958): 654–655.

[429] See M. Rosa, in *DBI* 3 (1961): 180–182. The Settele affair treated here and Anfossi's role in this connection, of course, escaped M. Rosa, as did other shorter works that Anfossi wrote during the course of it. They will be discussed at the proper place. A reference to the Settele Affair is found in an older short biography of Anfossi by R. Coulon, in *DHGE* 3 (1924): 1–2.

Synod of Pistoia which was so indebted to it. Anfossi treated the last-mentioned subject in two volumes. He dedicated two volumes to a refutation of the Gallican ideas, too, and he likewise grappled with the Civil Constitution of the Clergy in revolutionary France. Another subject was the divine foundation and indestructibility of the Church and of the See of Peter.[430] This work appeared in 1815. During that same year Pius VII appointed Anfossi, who was impeded in his activity during the French occupation, as Master of the Sacred Palace, after having assigned him the previous year together with Olivieri to be assistants to the ex-general and now general vicar of the Dominican order, Gaddi, who as a collaborator had fallen into disrepute.

The intransigence that Anfossi displayed now, especially in the question of the restitution of Church properties that had been seized over the course of the Napoleonic troubles, led him to clash with Pius VII and Leo XII, and with those around them. Predominant in his thinking, which relied entirely and decisively on Thomism, which was flourishing again,[431] were restorative, indeed reactionary features which — combined with a rather mediocre

[430] Some of the works by Anfossi (titles are translated from Italian): *Response of Fr. F. Anfossi, O.P. to the letters by Plat and to the opposition of some other theologians who claimed to contest the Bull* Auctorem fidei *in which the Synod of Pistoia is condemned by the Roman Pontiff Pius VI*, 2 vols. (Rome, 1805); an expanded new edition under the title *Defense of the Bull* Auctorem fidei *in which the major questions that have agitated the Church in these times are discussed*, 3 vols. (Rome, 1816) (in the appendix to vol. 3 we find Anfossi's "Considerations on the Civil Constitution of the Clergy and of the Constitutional Church of France"); *Reasons why Fr. F. A[nfossi], O.P., thought that it is not possible to adhere to the four Gallican propositions*, 2 vols. (Rome, 1813); *The divine institution and indefectibility of the Roman Catholic Church and of the See of Peter which is its basis* (Rome, 1815).

[431] Anfossi — like Olivieri — belonged to the Accademia di Religione Cattolica; he was elected a member of it on August 11, 1803. See Piolanti, *L'Accademia di Religione Cattolica*, 88.

intellectual talent, an honest, unselfish zeal for the Church's cause, and an unbending character — moved him to take a solitary stance in the Settele affair against the Holy Office, the Congregation of the Index, and indeed the pope himself, and against all of educated Rome. It may have been the experience of the Revolution and its aftermath that prejudiced him until his death on May 17, 1825, against any change at all of ideas and conditions and filled him in advance with skepticism about everything new. However, before we blame Anfossi on that account, we should acknowledge the selflessness of this resistance, by which he risked losing the goodwill of Pius VII and thus his employment, for the sake of his conviction.[432] We should not forget either that his obstinate resistance is precisely what gave the Holy Office the occasion for that convincing intellectual achievement, namely the argumentation that was to be presented forthwith in favor of admitting heliocentrism; without Anfossi, it would not have been necessary to produce it.

[432] Even when Anfossi was dead, Settele still conceded that he had acted out of noble motives. See Settele, April [*sic*; May] 20, 1825, in Maffei, *Giuseppe Settele*, 420.

CHAPTER 4
THE COURSE OF EVENTS

WHEN GIUSEPPE SETTELE had completed the manuscript for volume 2 of his *Elementi di Ottica e di Astronomia,* which contained astronomy, he turned to his colleague Olivieri with the question of whether in this book he could openly declare that the earth moves, without getting into trouble. Actually this inquiry is amazing, since even he must have known that long before him others had done this without eliciting the slightest reaction. Not surprisingly, therefore, the *compagno* of the commissioner of the Holy Office answered Settele's inquiry without hesitation and without qualification with a yes. It was all the more astonishing when Settele's publisher Filippo de Romanis[433] submitted the manuscript to the Master of the Sacred Palace, Fr. Anfossi, so that he could issue the imprimatur, and Anfossi forbade publication, citing the 1616 decree.[434] De Romanis objected that it was long since outmoded, but Anfossi replied by remarking that the Bible is still the same, and it teaches that the sun goes up and goes down, while the earth stands firm forever. In his diary, Settele commented on Anfossi's negative decision on January 3, 1820, with the remark that he was totally against freedom of the press, which was a plague on mankind.

[433] See Maria Iolanda Palazzolo, "Arcadi e Stampatori, I De Romanis nella Roma dei Papi," in *Tre secoli di storia dell'Arcadia,* 59–74, esp. 74 (with the interesting note that the publishing house de Romanis enjoyed special support from Anfossi!).

[434] See Settele, January 3, 1820, in Maffei, *Giuseppe Settele,* 285–286.

Censorship, though, should be entrusted to learned laymen and not to such stubborn mules as the *Frati* (the Dominican friars).

The reactions to Anfossi's attitude varied. Olivieri thought it best to bring the whole affair to the attention of the Holy Office for a thorough clarification,[435] while others, like Andrea Conti, a collaborator of the astronomer Calandrelli, suggested that Settele should only report the arguments in favor of the earth's motion, putting them on the lips of other authors.[436] Olivieri, however, did not think much of that solution, but began instead to collect material for the moment when the matter would be made dependent on the Holy Office after all.[437] He did well, because the pope was totally in agreement with that plan. But he balked at ordering Anfossi to issue the imprimatur, since he feared that the latter would protest loudly against it. So the rector of the Sapienza, Cristaldi, informed him.[438] Olivieri, on the other hand, had already figured out the main lines of his solution: it is possible to teach heliocentrism today without necessarily detracting thereby from the 1616 and 1633 decrees.[439] Whereas they had condemned the teaching that the earth moves as

435 See ibid., 289.

436 About Andrea Conti (1777–1840), see P. Maffei, in *DBI* 28 (1983): 347–348. Settele comments on Conti's recommendation (translated from Italian): "In short, we see that they are all sycophants, and deny the known truth so as not to run afoul of the persons who are in charge." Settele, January 10, 1820, in Maffei, *Giuseppe Settele*, 290.

437 See Settele, January 12, 1820, in Maffei, *Giuseppe Settele*, 291.

438 See Settele, January 17, 1820, in Maffei, *Giuseppe Settele*, 291. Belisario Cristaldi (1764–1831), jurist, known for his editions of Rota decisions, a consistorial advocate. He refused to take the oath to the regime of the French occupation. In 1817, he was appointed rector of the Sapienza. He did outstanding charitable work and was a friend of St. Gaspar del Bufalo. He stabilized the finances of the Papal States, which had had a deficit of 300,000 scudi. A close collaborator of Consalvi, he was made a cardinal in 1826 and ordained a priest in 1829. See M. Caffiero, in *DBI* 31 (1985): 1–4.

439 Brandmüller and Greipl, *Copernico, Galilei e la Chiesa*, 75.

philosophically absurd and heretical, we can no longer talk about absurdities since we know now about the centripetal force of gravity and the weight of the air, which precluded the absurd consequences of the earth's motion. Moreover, no one maintains now that the sun is the stationary center of the universe. Olivieri later on would dedicate himself to elaborating thoroughly the argument suggested here, with convincing results. It is not evidence of particular intelligence on Settele's part that he regarded this only as a sudden, opportunistic change of opinion by Olivieri.[440]

Although Olivieri clearly saw the intellectual approach to solving the problem, he and Msgr. Turiozzi were not sure about their way of proceeding.[441] So first they sounded out the intentions of the pope, to whom Msgr. Cristaldi presented Olivieri's arguments verbally.[442] The pontiff, however, obviously did not want to tie himself down and backed out of the affair with a quotation by Voltaire: When the philosopher spoke to Frederick II of Prussia about the proofs for the earth's movement, Frederick replied that the natives of Madagascar had proofs, too, for the movement of the sun. Thus both sides cited their experience.

Given that state of affairs, it was decided that Settele should request the imprimatur from the pope directly in writing. Olivieri drafted the petition, and after Turiozzi had approved it, Settele wrote

[440] "Fr. Olivieri is very persuasive about the Earth's movement, but he is a friar, and a Dominican, and belongs to the Inquisition, and that is enough to ruin his head" (translated from Italian). Settele, January 24, 1820, in Maffei, *Giuseppe Settele*, 293.

[441] The question was whether or not the Holy Office should attempt to deal with the affair. Settele, February 4, 1820, in Maffei, *Giuseppe Settele*, 294.

[442] Settele's commentary: "The Pope is a very cunning fellow [*un gran furbo*]." Settele, February 20, 1820, in Maffei, *Giuseppe Settele*, 295.

out in his own handwriting the six-page petition[443] and had it forwarded to the pope by Msgr. Soglia, a trusted papal advisor.[444] Plainly the intention was to induce the pope in this way to entrust the affair to the Holy Office, which he in fact did.[445]

Meanwhile, however, the Settele affair had taken on international dimensions. Whereas in mid-January it had already been the subject of lively commentary by the chattering classes of the city of Rome,[446] the foreign press immediately got a hold of it. The *Allgemeine Zeitung* in Augsburg, published by Johann Friedrich von Cotta, was the first to report on it. Cotta's correspondent in Rome was the Prussian consul-general Jakob Ludwig Salomo Bartholdy, who had held that office since 1815 and enjoyed the special confidence of Cardinal Secretary of State Consalvi.[447] Bartholdy had written a

443 See Settele, February 25 or March 1, 1820, in Maffei, *Giuseppe Settele,* 296–298. Text of the petition: ibid., 463–470.

444 Giovanni Soglia Ceroni (1779–1856) from Imola, an important jurist, closely associated with Pius VII from the time of his episcopal ministry in Imola, went with him into French exile, and experienced confinement there. Then professor at the Sapienza; in 1826 titular archbishop of Ephesus; in 1839, cardinal and bishop of Osimo and Cingoli, where he worked as an exemplary pastor until his death. A close friend of Rosmini's and of other important personages "*di tendenza moderata*" (of a moderate tendency), despite all his advantages he is said to have had an anxious character. See Piolanti, *L'Accademia di Religione Cattolica,* 48–49; Ritzler and Sefrin, *Hierarchia catholica,* vol. 7, 30, 188.

445 See Settele, March 17, 1820, in Maffei, *Giuseppe Settele,* 301. See also Bertazzoli's cover letter to Turiozzi dated March 19, 1820, in Brandmüller and Greipl, *Copernico, Galilei e la Chiesa,* 177–178.

446 Thus Settele, January 22, 1820, in Maffei, *Giuseppe Settele,* 292.

447 Jakob Ludwig Salomo Bartholdy (1779–1825), an uncle of the composer Felix Mendelssohn Bartholdy, also wrote a biography of Consalvi: J. L. S. Bartholdy, *Züge aus dem Leben des Cardinals Hercules Consalvi: Mit dessen Portrait* (Stuttgart: Cotta, 1825). See J. Kahl, "Johann Friedrich Cotta (1764–1832)," in Heinz-Dieter Fischer, ed., *Deutsche Presseverleger des 18. Bis 20. Jahrhunderts,* Publizistik-historische Beiträge 4 (Pullach: Verlag Dokumentation, 1975), 82–90; Eduard Heyck, *Die Allgemeine Zeitung*

report on the Settele affair dated February 6, and his report was published on March 2 in the *Allgemeine Zeitung*. On the previous day a somewhat abridged French translation, dated February 13, had appeared in the Parisian *Journal des Débats Politiques et Littéraires*.[448] This report is distinguished by its sober, objective, informative style, and its content was accurate, apart from the fact that it maintained that Benedict XIV had allowed the presentation of the Copernican system as a hypothesis, which as everyone knows was never forbidden. The article in the *Allgemeine Zeitung* was reprinted — again in abbreviated form — by the *Bayreuther Zeitung*, the *Bremer Zeitung*, the *Rheinische Blätter*, and the *Oppositions-Blatt/Weimarische Zeitung*.[449] In this way even Goethe certainly learned about it. So much for the repercussions in German-speaking lands.

Now, though, the reports in the *Journal des Débats* and the *Allgemeine Zeitung* became known in Rome around March 17 and were immediately the subject of intense debate. Some even voiced the suspicion that Settele himself had launched this communication, and Settele's friends, for instance Turiozzi and Cristaldi, thought that this publication was more likely to complicate the matter. Settele himself, however, had been surprised by it, and had

---

1798–1898: Beiträge zur Geschichte der deutschen Presse (Munich: Verlag der Allgemeine Zeitung, 1898), 140–141; and H. Haussherr, in *NDB* 1 (1953): 609. Considering the confidential relationship between Bartholdy and Consalvi, it is not improbable to assume that the cardinal secretary of state was informed about Bartholdy's intention to report about the Settele matter.

448 See the supplement to *Die Allgemeine Zeitung* (Augsburg), March 2, 1820, 125–126; *Journal des Débats Politiques et Littéraires* (Paris), March 1, 1820, 1.

449 See *Baireuther Zeitung* (Bayreuth), March 9, 1820, 200; supplement to the *Bremer Zeitung* (Bremen), March 11, 1820, no pagination; *Rheinische Blätter* (Wiesbaden), March 11, 1820, 163; *Oppositions-Blatt/Weimarische Zeitung* (Weimar), March 15, 1820, 511–512.

no trouble convincing Cristaldi of this. The Dutch ambassador was the one who informed him that Bartholdy was the author of the article.[450] However, this publication seems to have been less harmful than helpful, since it showed how urgent it had become to make a decision in favor of Settele.

In these circumstances Turiozzi, by order of the pope, turned to the secretary of the Congregation of the Index with the request to forward to him those files of the congregation that had been created in connection with the 1758 edition of the Index. Surely they would show what significance should be attributed to the fact that the general prohibition against writings which taught that the earth moves was no longer to be included in the new edition of the Index.

Father Bardani delivered the requested dossier on March 28, 1820.[451] Turiozzi immediately began studying the material; after all, with regard to the announcements in the press that could be read in Holland and Austria, too, he was convinced that the knot had to be untied quickly and elegantly now.[452] Plainly Turiozzi's intervention with Pius VII was successful,[453] because as early as April 14 Olivieri was able to inform the delighted Settele that the

[450] See Settele, March 18, 19, 20, 1820, in Maffei, *Giuseppe Settele*, 302–304.

[451] The text is in Brandmüller and Greipl, *Copernico, Galilei e la Chiesa*, 178–179; and Maffei, *Giuseppe Settele*, 533–534. It should be noted that Bardani writes "to set forth what I was able to find in the books of this Archive, and relative to what happened in the time of the Pontificate of Benedict XIV of blessed memory about the Authors and Books that deal with the Copernican system" (ibid., 533). This statement confirms the impression expressed in the introduction that the archival documents for all these procedures were minimal. In the present case there was only the briefest record of the result!

[452] See Settele, April 5, 1820, in Maffei, *Giuseppe Settele*, 305.

[453] Settele notes his intention to do this in his diary on April 3, 1820. Maffei, *Giuseppe Settele*, 305.

pope would now command Anfossi to issue the imprimatur.[454] Turiozzi and Bertazzoli, for their part, intended to influence Anfossi to this effect.[455]

So it happened. In order to communicate to Anfossi in due form the pope's decision, Turiozzi scheduled a separate appointment with him, about which Olivieri filed a report on May 13.[456] During this meeting — it does not say when it took place — Anfossi told Turiozzi that he obviously had great respect for the pope's orders, but nevertheless considered himself incapable of issuing the commanded imprimatur in his (Anfossi's) name. He took his stand with the teachings of the Church Fathers. Thereupon he showed Turiozzi an essay that he had written in which he criticized the foreign newspapers on account of their reportage in the Settele case.[457] It was a new edition of a publication in which Anfossi had grappled with the views of the Swiss natural philosopher Charles Bonnet.[458] In it he had taken the opportunity of the new edition, whatever its motivation may have been, to print a polemic against

[454] Olivieri tells Settele what Turiozzi had reported to him: "That the Pope will tell Fr. Master of the Sacred Palace to let my Astronomy with the earth's movement pass muster." Settele, April 14, 1820, in Maffei, *Giuseppe Settele*, 307.

[455] Soglia too probably worked along these lines, as his communication to Settele suggests: Settele, April 19, 1820, in Maffei, *Giuseppe Settele*, 308.

[456] See Settele, May 13, 1820, in Maffei, *Giuseppe Settele*, 313. Olivieri's report to the congregation is in Maffei, *Giuseppe Settele*, 432–433.

[457] See Settele, May 13, 1820, in Maffei, *Giuseppe Settele*, 313.

[458] About Charles Bonnet (1720–1793), see E. Behler, in *LThK2* 2 (1958): 601. Bonnet's work *La Palingénésie philosophoque* I-II (Geneva, 1769), had provoked Anfossi to write the following response: *Le fisiche rivoluzioni della natura o la Palingenesi filosofica di Carlo Bonnet convinta di errore: Dissertazione theologico-filosofica* (Venice, 1802; Rome, 1820). The polemic aimed against Settele is found in the second edition, 93–94 (reprinted also in Maffei, *Giuseppe Settele*, 536–537. For a synopsis of the content, see *Morgenblatt für gebildete Stände* (Stuttgart-Tübingen, November 20, 1820), 1116. Anfossi's biographer M. Rosa calls his work "a modest protest

Settele. Although as far as arguments against an imprimatur for his "astronomy" are concerned, he offers only a repetition of what had been said long ago, he nevertheless takes a stance against the aforementioned newspaper articles, whereby he refers to the *Journal des Débats* and not only repeats his reasons against an imprimatur for Settele, but also energetically boasts that granting permission to publish is his competence. In particular his ill humor is directed against the person who had launched the article that he criticized. Anfossi thought that Settele was the one responsible.[459]

Of course he had the misfortune of issuing by himself the imprimatur for his own publication —which earned him harsh criticism later on.[460]

Anfossi felt deeply affected by the pope's decision. He also sensed that he was becoming increasingly isolated. Even his official associate, Fr. Piazza, was now convinced by the argumentation of Settele and Olivieri and disavowed Anfossi.[461] By this time the latter did not hesitate to compile his reasons for refusing the imprimatur in a publication subdivided into nine points.[462]

Now it says a lot about Olivieri's character that under these circumstances he did not simply try to overlook Anfossi's resistance, but rather for his part also set about composing a publication with which he hoped to convince Anfossi.[463] Anfossi, however, in no way admitted defeat. Whatever his arguments may have been, he succeeded in winning over to his side the pope's majordomo, Msgr.

---

against the evolutionism of the Genevan naturalist, and more generally against Enlightenment materialism and optimism." *DBI* 3 (1961): 180).

459 See Settele, May 18, 1820, in Maffei, *Giuseppe Settele*, 313–314.

460 See Turiozzi to Anfossi, August 27, 1820, in Brandmüller and Greipl, *Copernico, Galilei e la Chiesa*, 306–307; Maffei, *Giuseppe Settele*, 546.

461 See Settele, May 15, 1820, in Maffei, *Giuseppe Settele*, 313.

462 For the text, see Maffei, *Giuseppe Settele*, 548–554.

463 See Settele, May 19, 1820, in Maffei, *Giuseppe Settele*, 314.

Antonio Frosini,[464] who now intervened with Turiozzi and Olivieri on behalf of Anfossi and suggested that Settele should change one point in his text and, instead of saying "*Movendosi la terra intorno al sole...*" (Since the earth moves around the sun...), should write "*Posto il moto della terra...*" (Assuming the earth's movement...).[465] Of course, with that they would have arrived again at the state of affairs from the year 1620.

Even the very communicative Settele tells us only a few scraps about what contacts took place between which dialogue partners in this situation and what the results were. Anfossi, in any case, did not remain silent: against the astronomer and physicist Ciccolini,[466] who had collaborated with Lalande in Paris, he tried to argue that the decrees of 1616 and 1633 remained valid. Anfossi seems once again to have impressed even the pope, who had wavered for a long time.

The majordomo, in contrast, still siding with the Master of the Sacred Palace, allowed Turiozzi to correct his misconception and now was also afraid that the Holy See could become a laughing stock, if Anfossi kept the upper hand.[467] This conviction spread more and

[464] Antonio Frosini (1751–1834) was from Modena and had an exclusively curial career. As "*Palatii Apostolici Praefectus*" (majordomo) he was created a cardinal in 1823. See Ritzler and Sefrin, *Hierarchia catholica*, vol. 7, 15. Settele notes the judgment of two colleagues about Frosini "that he is a man who understands nothing; who hates learned men; because he thinks that they are all unbelievers." Settele, June 27, 1820, in Maffei, *Giuseppe Settele*, 323.

[465] Settele, June 3, 1820, in Maffei, *Giuseppe Settele*, 318. The passage is from the text of Giuseppe Settele, *Elementi di Ottica e di Astronomia* II (Roma, 1819 [!]), 142.

[466] Ludovico Maria Ciccolini (1767–1854) belonged to the Order of St. Jerome. Astronomer and physicist, from 1798 to 1801 he was Lalande's collaborator in Paris. He was a professor in Bologna and from 1809 on authored many publications in *Connaissance des temps*. See L. Briatore, in *DBI* 25 (1981): 357–358.

[467] See Settele, June 17, 1820, in Maffei, *Giuseppe Settele*, 321.

more and increased the general annoyance at Anfossi, since his conduct was harming the Church.[468] By now, Turiozzi and Olivieri had arrived at the definitive conviction that the whole muddled affair could be dispatched only if the pope sent it to the Holy Office. However, before initiating steps for this purpose, they wanted to wait a few more days until Olivieri had been formally appointed commissioner of the Holy Office.[469]

THE HOLY OFFICE INTERVENES

No sooner was this done than Olivieri and Settele composed a new request in which Settele asked that his application for an imprimatur be referred to the Holy Office. When Turiozzi had approved the draft, too, Settele submitted his petition to Msgr. Soglia, who was supposed to forward it to the pope. Pius VII complied immediately with the request, and thus the Settele affair entered a new and decisive phase.[470] At the same time, Settele suggested to Olivieri that it was high time to finish the job and to strike Corpernicus and Galileo from the Index,

[468] A statement to this effect was made by an unnamed cardinal, who wanted to speak with Consalvi about it; Cardinal Guerrieri also thought so. Settele, July 1 and 3, 1820, in Maffei, *Giuseppe Settele*, 326–327. Anfossi's confrere Bardani thought that Anfossi was making the Dominicans even more ridiculous than they already were; everyone in the order who thought reasonably was in favor of Settele. Settele, July 11, 1820, in Maffei, *Giuseppe Settele*, 328–329. A few days later Settele was able to report on similar statements by the general superior of the Camillians, Fr. Michelangelo Toni, and by Msgr. Caprano. Settele, July 14, 1820, in Maffei, *Giuseppe Settele*, 329. Olivieri reported that Cardinal de Gregorio took Anfossi to task, "*gli fece una gran strillata*" (shouting loudly). Settele, July 18, 1820, in Maffei, *Giuseppe Settele*, 330. Cardinal Caccia-Piatti was the only one to tell Settele that he should not have been involved in the matter. Settele, July 27, 1820, in Maffei, *Giuseppe Settele*, 331.

[469] See Settele, July 18, 1820, in Maffei, *Giuseppe Settele*, 330.

[470] See Settele, August 1, 1820, in Maffei, *Giuseppe Settele*, 331–332. The pope granted the petition on August 2 (ibid., 332). Text of the petition: ibid., 535–538.

too.[471] Despite the fact that Rome was looking forward to the *Ferragosto* (Italian national holiday on August 15) and preparing for the annual trip *"ad aquas"* (to the waters, i.e., the coast), the staff of the Holy Office went to work without delay and energetically. On August 7 the question was on the agenda for the "*Feria Secunda*," in other words, the assembly of the consultors, who decided to delegate one of their group to prepare an expert opinion about how they might proceed and maintain the prestige of the Holy See.[472]

The cardinals approved this recommendation two days later in their "*Feria Quarta*,"[473] and they authorized the general procurator of the Barnabite order, Fr. Antonio Maria Grandi, to draw up this document.[474] After the return of Pius VII from exile, Grandi had been increasingly pressed into the service of the Curia. He was a consultor for several congregations and since 1819 secretary of the Congregation for Extraordinary Ecclesiastical Affairs. In this last-mentioned capacity he was one of Consalvi's close collaborators. Grandi got to

[471] See Settele, August 4, 1820, in Maffei, *Giuseppe Settele*, 332, 335.

[472] Only one of the eleven consultors was of the opinion that they all should express their views about it. See minutes of the meeting on August 7, 1820, in Brandmüller and Greipl, *Copernico, Galilei e la Chiesa*, 293.

[473] See minutes of the meeting on August 9, 1820, in Brandmüller and Greipl, *Copernico, Galilei e la Chiesa*, 298.

[474] Antonio Maria Grandi (1760–1822) was from Vicenza, earned a doctorate in theology in Pavia, and taught in Milan, Cremona, and Bologna. After the election of Pius VII he held high offices in the Barnabite order and served as a consultor for various congregations in Rome; he became involved in the Accademia di Religione Cattolica. Although less outstanding for his erudition, he was impressive because of his comprehensive education, especially of a literary sort. He was also known as a poet. See Antonio Cesari, *Per la morte del P. Antonio Grandi Vicentino, Vicario Generale de'Barnabiti in Roma* (Verona: Pietro Bisesti, 1822); Giovanni Piantoni, *Elogio storico al R. P. Don Antonmaria Grandi barnabita* (Rome: Stamperia della S. Congregazione de Prop. Fide, 1858); Moroni, *Dizionario*, vol. 16 (1842), 157–158; Piolanti, *L'Accademia di Religione Cattolica*, index; Boffito, *Scrittori Barnabiti*, 266–272; L. Pásztor, in *DHGE* 21 (1986): 1113–1114.

work immediately and by August 12 already submitted to Olivieri a draft with the request for suggestions about how to revise it, if needed.[475] Two days later the consultors decided to have copies made of the expert opinion to present to all the cardinals of the congregation.[476] The stated purpose of the opinion was therefore to explain how the Holy See could acknowledge the teaching about the earth's movement, which had been condemned in 1616, without damaging its prestige.[477] Now Grandi first points out the fact that as early as 1620 the Congregation of the Index explicitly allowed advocating the Corpernican system as a hypothesis. It is clear from this that in 1616 it could not have been condemned at all as an error of faith or a heresy, for such things could not be admitted even in the form of a hypothesis, since a hypothesis might someday be proved true. In fact at that time a corresponding vote by the qualifactors of the congregation had not been taken. Instead they had limited themselves to condemning Galileo's teaching as harmful and contrary to Sacred Scripture. This, however, referred to the literal sense of Sacred Scripture. Baronius, however, already remarked during Galileo's lifetime: *"Spiritui Sancto mentem fuisse nos docere, quomodo ad Coelum eatur, non quomodo Coelum gradiatur"* (The Holy Spirit intended to teach us how to go to Heaven, not how the heavens go).[478]

Grandi argued that, although in 1664 the two Index decrees from 1616 and 1620 were included in the Index in the list of the other decrees, nevertheless they did not appear in the following editions of the Index either verbatim or in the form of a general

[475] See Grandi to Olivieri, August 12, 1820, in Brandmüller and Greipl, *Copernico, Galilei e la Chiesa*, 298.

[476] See Olivieri's note to the file dated August 14, 1820, in Brandmüller and Greipl, *Copernico, Galilei e la Chiesa*, 299; Maffei, *Giuseppe Settele*, 542.

[477] The text of the expert opinion is in Brandmüller and Greipl, *Copernico, Galilei e la Chiesa*, 294–298; and Maffei, *Giuseppe Settele*, 539–542.

[478] Brandmüller and Greipl, *Copernico, Galilei e la Chiesa*, 295.

Index rule. This is of course a statement that is easily refuted by inspection. Every edition of the Index from 1619 until 1758 contained the rule that *"Libri omnes docentes mobilitatem terrae et immobilitatem solis"* (All books teaching that the earth moves and the sun stands still) were forbidden. How both Grandi and also Olivieri could come to the opposite conclusion is not evident.

Therefore, the report continued, the Copernican system won more and more adherents, since its weaknesses had been overcome and new proofs had been brought to light. Grandi cites his learned confrere, Cardinal Gerdil, who in his book *Storia delle Sette de'Filosofi* had expressed his admiration of Copernicus.[479] He also quotes the latest astronomical publications by Guglielmini and Calandrelli; the latter had even been able to dedicate his *Opuscoli* to Pius VII. All these facts deprived the former prohibitions of all foundation, which is why, in his opinion, relevant publications could not be obstructed.

But how, he continues, should we proceed in practice, if we do not want to offend the congregations involved or the Master of the Sacred Palace? Certainly the latter's zeal for the purity of the faith is praiseworthy, but a judgment in this matter — given the ever wider extension of all fields of knowledge — goes beyond the limits of his competence. He, Grandi, was convinced that Anfossi was acquainted neither with the relevant files of the Congregation of the Index nor with the imprimaturs that had been issued in recent times. His recommendation therefore reads, *"Nihil obstare, quominus defendi possit sententia Copernici de motu telluris, eo modo, quo*

[479] Giacinto Sigismond Gerdil's *Storia delle Sette de'Filosofi* (History of the Philosophers' Sects) was included in the edition of his collected writings that was published in Rome from 1806 on: *Opere edite ed inedite del Cardinale S. Gerdil* vols. 1–20 (Rome, 1806–1821); the reference here is to 1:258–259. Grandi participated in editing the works of his confrere Gerdil. See Hurter, *Nomenclator Literarius Theologiae Catholicae,* 861.

nunc ab Auctoribus etiam Catholicis defendi solet" (Nothing hinders [the author] from defending the opinion of Copernicus about the movement of the earth, in the way in which it is usually defended now by Catholic authors).[480] Therefore it should be suggested (*insinuetur*) to Anfossi that he not hinder the publication of Settele's work. It should be recommended to Settele, however, that he show in an addition to his manuscript that the Copernican system in its present form is no longer burdened by those difficulties under which it labored before the later discoveries.

As early as August 16 the congregation adopted this recommendation of its consultor, indeed unanimously. The pope immediately confirmed the decision.[481] Thus they could proceed to execute it. On the following day, therefore, Turiozzi recommended that the required addition to Settele's work should be formulated in collaboration with Olivieri and Grandi, whereby they could cite Cagnoli's *Notizie astronomiche,* an excerpt from which Turiozzi likewise presented.[482]

Two days later, when Settele brought his draft, Grandi was extremely critical of his excessively Scholastic language, and opined that many questions should not be touched on in the first place, and

480 Maffei, *Giuseppe Settele,* 542.

481 See the minutes to the meeting on August 16, 1820, in Brandmüller and Greipl, *Copernico, Galilei e la Chiesa,* 299–300; Favaro, *Le Opere di G. Galilei,* 19:419). The text of the recommendation that was raised to the status of a decision is in Brandmüller and Greipl, *Copernico, Galilei e la Chiesa,* 300–301.

482 See Settele, August 17, 1820, in Maffei, *Giuseppe Settele,* 337–339. The text of the excerpt is in Brandmüller and Greipl, *Copernico, Galilei e la Chiesa,* 302–303. Antonio Cagnoli (1743–1816) was an astronomer, friend of Lalande, from 1785 on again in Italy, 1796 president of the Italian Society for the Progress of Science, professor of mathematics at the military academy in Modena. His *Almanacco con diverse notizie astronomiche adattate all'uso comune* was published from 1787 to 1806 in sixteen volumes in Modena. See U. Baldini, in *DBI* 16 (1973): 325–327.

that furthermore he should limit himself to citing those statements from Cagnoli and Gerdil in which they mention Copernicus approvingly. Settele, however, was not at all angry at him for this, but rather lavished praise on Grandi in his diary on account of his mathematical and literary knowledge.[483] Now Grandi's recommendations were discussed with Olivieri, too, and Settele went to work again; the result pleased both Grandi and also Olivieri.[484] Just to make sure, Settele presented his draft on August 22 to Cardinal della Somaglia, too. The cardinal found it very good, even called Settele "another Galileo," and furthermore amazed him by discussing fixed star parallax with him quite knowledgeably.[485] The following day the cardinals of the congregation approved the text of Settele's *Nota,* and on August 26 he brought his manuscript to the printer.[486]

In order to prevent renewed interventions by Anfossi, the congregation dispatched together with Settele's *Nota* the draft of a letter to the Master of the Sacred Palace,[487] which was delivered to him on August 27. In this letter the assessor informed Anfossi that Settele had obtained from the pope an examination by the Holy Office of his work in which he advocated the movement of the earth and the immobility of the sun, not as a hypothesis.[488] After examining all the relevant proceedings from the past, the congregation, in keeping with the vote of the consultors, had decided "*che*

483 See Settele, August 19, 1820, in Maffei, *Giuseppe Settele,* 339.

484 See Settele, August 20, 1820, in ibid., 340–341.

485 See Settele, August 22, 1820, in ibid., 341–343.

486 See Settele, August 26, 1820, in ibid., 345. The text of the *Nota* is in Brandmüller and Greipl, *Copernico, Galilei e la Chiesa,* 303–305.

487 See minutes of the meeting on August 23, 1820, in Brandmüller and Greipl, *Copernico, Galilei e la Chiesa,* 303.

488 See Turiozzi to Anfossi, August 27, 1820, in Brandmüller and Greipl, *Copernico, Galilei e la Chiesa,* 306–307; Settele, August 27, 1820, in Maffei, *Giuseppe Settele,* 345–346.

nulla osta, che si difenda la Tesi della mobilità della Terra, e immobilità del Sole nel modo, che comunemente in oggi s'insegna dagli autori Cattolici."[489] The cardinals had therefore ordered that he, Anfossi, be informed that he could "*con sicurezza*" (surely) allow Settele's work to be printed and made public, all the more so because they had made sure that the work proved that the reasons for the earlier prohibitions no longer existed. The cardinals had also ordered him, Turiozzi, to present all this to the pope and also to ask him to remind Anfossi that the Council of Trent forbade all members of religious orders, and therefore also the Master of the Sacred Palace, to publish their own works without the permission of their religious superiors. Moreover, he was not allowed to issue by himself the imprimatur for his own works — the vicariate must do this. Finally, in the name of the pope complete silence with regard to the Settele affair should be imposed on him. Per instructions, Turiozzi presented all this to the pope, who granted his approval for it. Therefore Anfossi was requested to behave accordingly.[490]

Anfossi reacted promptly to these communications, which understandably offended him deeply. On August 28 he issued the imprimatur for the book with the *Nota* — as de Romanis immediately wrote to Settele.[491] In a somewhat more thorough letter he turned to Turiozzi also and, while repeating his fundamental reservations, also solemnly declared that he subjected his own judgment to

[489] For an English translation, see footnote 50. This version replaces "is defended" with "is taught." Brandmüller and Greipl, *Copernico, Galilei e la Chiesa*, 306.

[490] Settele learned from his (secular) colleague Domenico Morichini that the pope rebuked Anfossi sharply in a letter and in doing so also criticized the fact that he had issued an imprimatur for his own publication. Settele, August 28, 1820, in Maffei, *Giuseppe Settele*, 347.

[491] Ibid.

the pope's.[492] He merely asked Turiozzi to present to him one or two more relevant opinions about Settele's work, the authors of which Turiozzi himself could decide. The imprimatur for Settele was now unavoidable, yet Anfossi seems to have been far more upset about the reference to those who were competent for the imprimatur of his own publications, for he dedicated to this question the lion's share of the letter in which he defended his conduct with good reasons.

New obstacles

By human reckoning, the Settele-Anfossi file should have been closed with the pope's decision. However, that did not happen. First Turiozzi complied with Anfossi's demand and did in fact commission two expert opinions. To compose them he turned to two personages of indisputable intellectual caliber with curial experience. It had long since ceased to be a matter of astronomy! One of them was Msgr. Pietro Ostini.[493] He was born in Rome in 1775, and received a doctorate in theology there in 1796, and priestly ordination two years later. As professor of Church history at the Roman College he gathered around himself an illustrious circle of artists, literati, and scientists from Protestant lands, quite a few of whom he accompanied on their journey into the Catholic

[492] See Anfossi to Turiozzi, August 28, 1820, in Brandmüller and Greipl, *Copernico, Galilei e la Chiesa*, 307–310; and Settele, August 29, 1820, in Maffei, *Giuseppe Settele*, 347.

[493] See Ritzler and Sefrin, *Hierarchia catholica*, vol. 7, 28, 360; Piolanti, *L'Accademia di Religione Cattolica*, 104; Schmidlin, *Papstgeschichte*, 268 and index; Moroni, *Dizionario*, vol. 50 (1851), 56–57. Ostini was the spiritual director for converts such as Ferdinand von Eckstein (1790–1861, conversion 1807), Friedrich Ludwig Zacharias Werner (1768–1823, conversion 1810), and Johann Friedrich Overbeck (1789–1869, conversion 1813), together with his friends Rudolf and Wilhelm Schadow, Karl Vogel von Vogelstein, Christian Schlosser, Friedrich von Hurter (1787–1865, conversion 1844). See David August Rosenthal, *Convertitenbilder aus dem neunzehnten Jahrhundert* 1/1 (Schaffhausen, 1871), 71, 182, 195, 260, 318; 1/2 (Schaffhausen, 1871), 280.

Church. Eminent among them is Karl Ludwig von Haller, who converted in approximately 1818. Later Ostini was active as a diplomat, in Switzerland and elsewhere, and as a nuncio in Vienna; in 1831 named cardinal *in pectore*; his status was declared in 1836, and he died in 1849 in Naples.

The other person approached for an expert opinion was the Carmelite Fr. Giuseppe Maria Mazzetti.[494] He was from Chieti, where he was born in 1778. He was first a physician, then a priest, and finally a Carmelite. Besides holding high-ranking offices in his order, he also performed curial functions as a consultor for many congregations. In 1821 he became secretary of the Accademia di Religione Cattolica, in 1836 bishop of Sora, Aquino, and Pontecorvo, and two years later titular archbishop of Seleucia, since meanwhile, as rector of the University of Naples, he had very successfully carried out reforms of the instructional and educational system in the royal service. Mazzetti died in 1850 in Naples.

Both experts completed their task within a few days. Both affirmed that Settele's work was at no point contrary to the Catholic faith or good morals, but rather excelled by its great scientific and didactic merits.[495] While this work was still under way, Anfossi intervened once again by having the manuscript of the *Nota* that had been required of Settele confiscated at the print shop, whereby he brushed

[494] Mazzetti himself wrote *Elementi di prospettiva lineare* (Rome, 1830); *Progetto di riforme pel regolamento della pubblica istruzione* (Naples, 1841). See Ritzler and Sefrin, *Hierarchia catholica*, vol. 7, 84, 341; Piolanti, *L'Accademia di Religione Cattolica*, 52; V. Apreda, *Elogio funebre di Msgr. G. Mazzetti* (Naples, 1850).

[495] See Mazzetti to the Holy Office, September 5, 1820, and Ostini to the Holy Office, September 12, 1820, in Brandmüller and Greipl, *Copernico, Galilei e la Chiesa*, 327 and 329. See also Turiozzi's cover letter dated September 14, 1820, in Brandmüller and Greipl, *Copernico, Galilei e la Chiesa*, 330–331.

off the objections of de Romanis by claiming that the pope had been hoodwinked and for that reason alone had given his approval for the printing. Therefore, he had to speak about it first with the pope, and the printing must not start until then.[496] Informed immediately by Settele about these procedures, Turiozzi decided to ignore this new intervention by the Master of the Sacred Palace completely, and instead to wait for the reviewers' opinions that he had demanded.

When Turiozzi forwarded these two opinions to Anfossi on September 14, his conditions for issuing the imprimatur were thereby fulfilled. Nevertheless, three days later he received Anfossi's laconic and no longer altogether surprising message that, having personally examined the additional *Nota*, he now revoked the imprimatur for Settele's work, which he had issued on the assumption that it was in keeping with the pope's will. He wanted nothing to do with a matter that damaged the prestige of the Holy See and was a violation of his duties. He had communicated his reasons to personages who were beyond all doubt reliable, and had received their consent. With that, the matter was for him definitively settled.[497] De Romanis too received a communication with similar contents.[498]

Now Anfossi probably felt after all that his action was too abrupt, for he turned again to Turiozzi with a longer letter, in which he tried to justify his refusal.[499] Settele was the one, he wrote, who had been unwilling to comply with his suggested formulation, which would have dispelled all misgivings. Of course this did not surprise him, for Settele in his day was among those who had taken the oath

[496] See Settele, September 2, 1820, in Maffei, *Giuseppe Settele*, 350.

[497] See Anfossi to Turiozzi, September 17, 1820, in Brandmüller and Greipl, *Copernico, Galilei e la Chiesa*, 331.

[498] See Settele, September 21, 1820, in Maffei, *Giuseppe Settele*, 357–358.

[499] See Anfossi to Turiozzi, September 22, 1820, in Brandmüller and Greipl, *Copernico, Galilei e la Chiesa*, 332–333.

to the French occupational regime. This was in fact true. In connection with the imprimatur for an astronomical textbook, and given the fact that this oath was now almost ten years in the past, this accusation cannot be called anything but irrelevant and unjustified. Nevertheless it certainly indicates a deep, emotional root of Anfossi's aversion to the *collaborateur* Settele, whereas he himself had remained loyal and had been persecuted by the French. Besides, Anfossi concludes, he left all further arrangements to Turiozzi. Moreover, he set about composing his *Ragioni* (Reasons), a fifteen-page report that he sent to the printer and delivered to the pope on October 7.[500]

Such letters and such conduct could not remain without an appropriate response, and it is not surprising that Turiozzi freely vented his anger—in very polite language.[501] At the conclusion he says: Doesn't Anfossi's conduct show quite clearly that he values his own personal theological judgment much more than that of highly respectable specialists, of all the consultors, indeed, of all the cardinal-inquisitors? The most reverend Father in his wisdom will surely be able to gauge how modest this self-assessment is.

In this same letter Turiozzi also wrote that the issuing of the imprimatur was now no longer Anfossi's business but rather that of the Holy Office. That was true. In addition, on September 20 the

[500] See Brandmüller and Greipl, *Copernico, Galilei e la Chiesa*, 336–349. It cannot be determined when Anfossi composed his seventy-six-page anonymous work: *Se possa difendersi ed insegnare non come semplice ipotesi ma come verissima e come tesi la mobilità della terra, e la stabilità del Sole da chi ha fatta la professione di Fede di Pio IV, Questione teologico-morale* (Rome, 1822). Olivieri says that this work, directed against Settele and him, was printed in the summer of 1822. See Settele, July 26, 1822, in Maffei, *Giuseppe* Settele, 411.

[501] See Turiozzi to Anfossi, September 22, 1820, in Brandmüller and Greipl, *Copernico, Galilei e la Chiesa*, 333–335.

pope had ordered the printing of Settele's book. "But who should give the imprimatur now?" Turiozzi had asked him. The pope had replied that he should just print it, that he would find someone soon enough to sign the imprimatur.[502]

But it was not to be all that easy, and there was still no end of the complications! Then, too, the press had taken up the case once more.

Again the *Allgemeine Zeitung* in Augsburg was the newspaper in which subscribers could read on September 7 a report dated August 26 about the positive decision of the Holy Office concerning Settele which meanwhile had been made.[503] The *Rheinische Blätter*, the *Augsburger Politische Abendzeitung*, and the *Oppositions-Blatt* reprinted this news.[504] Meanwhile de Romanis had begun printing the work, while in the Holy Office the question of who should now sign the imprimatur was discussed.[505] Anfossi could not acquiesce in this course of events, which was becoming more and more disadvantageous for him. Therefore, he wrote a treatise entitled *Ragioni per cui il P. Maestro del S. Palazzo Apostolico ha creduto e crede che non si può permettere la Stampa del Manoscritto del Signor Canonico Settele, che incomincia: Movendosi la Terra intorno al Sole* (Reasons why the Master of the Sacred Palace believed and believes that it is not possible to

[502] See Settele, September 22, 1820, in Maffei, *Giuseppe Settele*, 358–359.

[503] See *Allgemeine Zeitung* (Augsburg), September 7, 1820, 1003 (reprinted in Italian translation in Brandmüller and Greipl, *Copernico, Galilei e la Chiesa*, 328–329).

[504] See *Rheinische Blätter* (Wiesbaden), September 12, 1820, 591 (verbatim reprint from the *Allgemeine Zeitung*); *Augsburger Politische Abendzeitung* (Augsburg), September 15, 1820, 897 (verbatim reprint from the *Allgemeine Zeitung*); *Oppositions-Blatt/Weimarische Zeitung* (Weimar), September 18, 1820, 1775: "The Holy Office in Rome has finally admitted that the earth revolves around the sun, and allowed the Copernican cosmology to be presented publicly."

[505] See Settele, October 5–10, 1820, in Maffei, *Giuseppe Settele*, 361–364.

allow the printing of the manuscript by Canon Settele which begins: Since the Earth moves around the Sun).[506]

In early October he submitted it in print to the pope, who on October 11 had his majordomo Antonio Frosini forward it to the Holy Office. Frosini's cover letter shows that Anfossi, with his *Ragioni,* had evidently succeeded in gaining ground.[507] He accuses Settele of impertinence, since he did not hesitate to pit the Master of the Sacred Palace and the Holy Office against each other, whereas a single word — *ipotesi* (hypothesis) — could have done away with all the difficulties. To agree to do that is, after all, really not a sacrifice or an expression of religious prejudice or bigotry, Frosini insisted. Indeed, one can affirm as certain only what is true or what one absolutely believes. Can Settele believe with certainty that the Copernican system is absolutely true? To maintain this would be rash, after all! And now the classic recourse to Thomas Aquinas, although he is not mentioned by name: In order to characterize the Copernican system legitimately as absolutely true, it would be necessary "*che sapesse tutte le risorse possibili nel sistema planetario capaci di supplire al supposto movimento della terra, ed al riposo centrale del Sole*" (to know all the possible options of the planetary system capable of producing the supposed movement of the earth and the central rest of the Sun).[508] That, however, surpasses human capabilities. Frosini is, in any case, "*convinto di tutta l'incertezza, e della molta impostura della scienza Astronomica*" (convinced of the complete uncertainty and of the great imposture of Astronomical science).[509] Was it now the voice of the pope that had become uncertain?

506 The text is in Brandmüller and Greipl, *Copernico, Galilei e la Chiesa,* 336–349. The pope forbade the publication of the *Ragioni,* as Olivieri informed Settele. Settele, January 5, 1821, in Maffei, *Giuseppe Settele,* 386.

507 Frosini's cover letter, dated October 11, 1820, in Brandmüller and Greipl, *Copernico, Galilei e la Chiesa,* 349–350.

508 Ibid.

509 Ibid.

On the very next day Settele received an invitation from the Dominican Giuseppe Vincenzo Airenti, who a few days earlier had been consecrated bishop of Savona by Cardinal della Somaglia. Airenti, who belonged to the same Lombardy province of the order as Olivieri and Anfossi, plainly was in close contact with the latter.[510] Now he tried to influence Settele along these lines. Olivieri and Turiozzi, in contrast, confirmed him. However, everything was "in suspense" again, since it was feared that the pope might change his mind due to the influence of Settele's opponents. Moreover, there were increasing indications that Anfossi was about to win over additional allies.[511] In fact Settele — by now the date was October 23 — found upon returning from an errand a letter from the Papal Almoner Bertazzoli asking him for an interview. Settele, who suspected fresh complications, immediately consulted with Olivieri, who advised him that, come what may, he could always rely on the Holy Office.

Settele's fears were not unfounded, for Bertazzoli tried over the course of an interview lasting a whole hour to convert him to the term "hypothesis," whereby he cited a change of mind by the pope.[512] Settele nevertheless stood firm, and he did not fail to point out that, given the latest press reports, the prestige of Rome was at stake.

[510] See Settele, October 13, 1820, in Maffei, *Giuseppe Settele*, 364. Giuseppe Antonio (name in religion: Giuseppe Vincenzo) Airenti (1767–1831), after teaching and administrative duties in various localities, became the librarian of the Casanatense, in 1816 "*Teologo Casanatense*," and in 1820 bishop of Savona. He authored works about geographical, chemical, philological, and historical subjects. See G. Oreste, in *DBI* 1 (1960): 537–538; Ritzler and Sefrin, *Hierarchia catholica*, vol. 7, 334.

[511] Olivieri had informed Settele — probably on October 21 — that the printing should now be concluded and afterward the imprimatur granted. At least thirty to forty copies were supposed to be completed. This made room for conjectures about a new turn of events. Settele, October 21, 1820, in Maffei, *Giuseppe Settele*, 366.

[512] For a more thorough report about the conversation, see Settele, October 23, 1820, in ibid., 366–368.

Olivieri, to whom Settele reported everything immediately, advised him to point out, in reply to any mention of an order by the pope, that the pope ordered the printing through the Holy Office and, should he now have decided otherwise, should also communicate this to him through the Holy Office.[513]

In another conversation that Settele had on October 28 with Bertazzoli, he succeeded in convincing the almoner that it would not do for the pope to say in a private consultation something different from what he had said officially to the Holy Office. It could scarcely be assumed that the entire Holy Office conspired to deceive the pope. These arguments made such an impression on Bertazzoli that he declared himself willing now to present to the pope not only Settele's *Nota* but also the latest article of the *Allgemeine Zeitung*.[514] This seemed all the more appropriate since on October 25 the *Gazzetta di Genova* too, via the *Wiener Zeitung*, had reprinted a summary of the article.[515] A few days before that the *Morgenblatt für gebildete Stände*, which appeared in Stuttgart and Tübingen and was likewise published by Cotta, had reported under the dateline "September 30, Rome," that Anfossi, despite the decision of the highest authorities, had by no means given up his opposition to the publication of Settele's book.[516]

New expert opinions — who issues the imprimatur?

There still could be no talk about a forthcoming end to the conflict; on the contrary, in the Holy Office they feared additional, serious

[513] See Settele, October 23, 1820, in ibid., 368.

[514] See Settele, October 28, 1820, in ibid., 369–370. The article in question was the one mentioned in footnote 69, dated September 7, 1820.

[515] See *Wiener Zeitung*, September 16, 1820, 841; *Gazzetta di Genova*, October 25, 1820, no pagination (reprinted also in Maffei, *Giuseppe Settele*, 562).

[516] See *Morgenblatt für gebildete Stände* (Stuttgart-Tübingen), October 19, 1820, 1008.

obstacles to the printing of Settele's *Astronomia*. This is demonstrated above all by the fact that Anfossi's *Ragioni* did not fail to have its effect, although it was only the repetition of arguments that had long since been adduced.[517] At the same time as Olivieri, another consultor, Fr. Antonio Maria Grandi, once again was dealing with the affair. He wrote a seventeen-page expert opinion, the shrewdness of which cannot be denied.[518] We learn from it moreover the exact formulation of the questions that were presented to the consultation on November 20, 1820: (1) Whether any measure should be taken against the Master of the Sacred Palace, and if so, what? (2) Whether, given the arguments that Anfossi had submitted to the pope, the Holy Office should disavow the decision that it had made?

Fr. Grandi, too, harshly criticized Anfossi's incomprehensible conduct. Given the overall attitude of the Holy See toward the teachings of Copernicus and Galileo since 1620, such behavior must be called inappropriate and harmful, indeed irresponsible, he averred. Anfossi's opposition to the decision of the Holy Office dated August 16 is unacceptable. Now the assessor, Msgr. Turiozzi, must be commissioned to ask the pope himself to confront Anfossi.

Fr. Grandi's remarks on the second question are much more extensive and also, because of his shrewd argumentation, much more interesting. It is not surprising that he answers it in the negative. And these are his reasons: Even though the twofold movement of the earth, as astronomers teach it today, cannot be proved strictly and mathematically, the possibility remains nevertheless of conducting the proof by methods from physics. If someone is unwilling to acknowledge even that, he nevertheless cannot deny that this view has attained the highest degree of probability. If the Holy See

[517] See Settele, November 5, 1820, in Maffei, *Giuseppe Settele*, 371.

[518] Brandmüller and Greipl, *Copernico, Galilei e la Chiesa*, 386–393.

nonetheless were to insist on its position from the year 1616, the enemies of the Church, who were watching her behavior like a hawk, would make loud public outcry, as Voltaire had once done, and attack her on account of the ignorance with which she hinders the progress of science.

Above all, however, it must be considered that back then the Church condemned Galileo's theory as false and contrary to Sacred Scripture, which with regard to the literal meaning of Sacred Scripture was also justified. This literal meaning must not be abandoned, even when applying all the hermeneutical rules, unless it leads to absurd conclusions. But both Copernicus and Galileo presented their system in a form that was fraught with major difficulties, without resolving them satisfactorily, since they did not yet know about the weight of the air. For this reason, Grandi notes, the Holy Office at that time insisted on the literal interpretation of Sacred Scripture.

Considering the later discoveries, however, there is no longer any reason to hold this position. Instead, it is necessary from now on to interpret Sacred Scripture "*in senso figurato*" (figuratively). Anfossi's arguments do not change this in the least. Yet he also cites the irrevocability of papal decisions, which applies also to the Galileo case. Here, though, Grandi asserts, Anfossi ignores the fact that this irrevocable character belongs only to *ex cathedra* decisions, but by no means to decisions of the sort that were made in this case. Therefore, it is necessary to stay with the position taken by the Holy See, not to hinder the printing of books that teach the Copernican system. Reviewing the decision dated August 16, 1820, is out of the question. Particularly interesting, nevertheless, is Grandi's opinion of the oft-mentioned *Nota* by Settele. Grandi writes that it deserves the utmost attention, since after all it considers the "*assurdi filosofici*" (philosophical absurdities) inherent in the Copernican theory as the reason for its condemnation.

Grandi rejects this argument. He points out instead that the Holy See in magisterial decisions has never taken its orientation from philosophical criteria but rather from divine revelation alone, as it is available in Scripture and Tradition. Galileo was not condemned because he presented a false and philosophically absurd teaching, which therefore was also contrary to Sacred Scripture, but rather on account of the suspicion of heresy that he had incurred by the fact that he advocated a view that contradicted Sacred Scripture.

Now, since no reference is made here to philosophical errors, Grandi continues, certainly no one can maintain that such errors were the reason for the condemnation. Moreover, the objectionable "absurdities" were by no means the content of Galileo's teaching; rather, the censors had drawn these absurd consequences from Galileo's theory and, not knowing that air has weight, he was unable to refute them at that time. Therefore they could not be considered as the reason for the condemnation.

To this Grandi adds a momentous observation: it is very risky to make detailed statements about the more particular circumstances of Galileo's condemnation, since the acts of the trial had not yet been brought back from Paris and therefore were inaccessible.

From all this Grandi draws the conclusion that Settele's *Nota* in its present form cannot be approved. On the other hand, though, given the Curia's position so far in this affair, he sees no need either of adding anything at all to Settele's book. Therefore, Settele's work should be printed without any further explanation.

Olivieri therefore now felt obliged to compile the archival material available so far on the well-known *Positio* and to have it printed for official use. Moreover this was done hastily, as shown by the large number of typographical errors that crept in.[519]

[519] See ibid., 351–379. Anfossi reacted to it with his *Breve risposta* without adducing any new argument in it (ibid., 380–384).

On November 16 he showed Settele the printed dossier, while obliging him to the strictest confidentiality concerning it.[520] For all the exchange of arguments pro and contra, the quotations from Fr. Boscovich, which Anfossi repeated ad nauseam, seem to have played a particular role. After Olivieri had spoken about it with Caprano, too, the latter provided him with a more detailed analysis of the text passages in question, whereby he was able to demonstrate convincingly that Anfossi had twisted the meaning of Boscovich's statements into their opposite by the well-established method of taking individual sentences out of context.[521]

The *Consulta* of the Holy Office dealt with all this in their *Feria Secunda* meeting on November 20, 1820, which however was in no way impressed by the resistance that had come to light.[522] Ten of the consultors pleaded in favor of the immediate printing of Settele's work together with the *Nota* that had been required. One went even farther and recommended closing the files definitively. The imprimatur should be given by Cardinal della Somaglia as secretary of the Holy Office. This should be communicated to the majordomo Frosini, whereby they should also give expression to the astonishment of the Holy Office at his activities.

In fact the appearance of Settele's *Astronomia* was by now unstoppable.[523] On December 14 the assembly of the cardinals of the Holy Office decided that the vice-regent, Archbishop Frattini, should issue the imprimatur, after the two reviewers had given their

520 See Settele, November 16, 1820, in Maffei, *Giuseppe Settele*, 374.

521 See Caprano to Olivieri, November 18, 1820, in Brandmüller and Greipl, *Copernico, Galilei e la Chiesa*, 385–386.

522 See the minutes of the meeting on November 20, 1820, in ibid., 394.

523 At Olivieri's advice, de Romanis had resumed the printing. Settele, November 27, 1820, in Maffei, *Giuseppe Settele*, 376. See also the *Mogenblatt für gebildete Stände* (Stuttgart-Tübingen), December 5, 1820, 1168; December 11, 1820, 1188.

approval.[524] The assessor, Msgr. Turiozzi, on the other hand, should ask the pope to impose silence in this matter on both Anfossi and the majordomo Frosini. However, the decision does not seem to have been easy for the pope.[525] In Olivieri's opinion, "*non ha petto fermo*" (he has no intestinal fortitude).[526]

On December 21 matters had progressed to the point where Olivieri could tell Settele that the decision was made and the vice-regent was to issue the imprimatur. The publication of the book was to follow without causing a sensation — which certainly was not to Settele's way of thinking.[527] To all appearances, in the Holy Office they feared further difficulties, despite everything, for Turiozzi was now urging a swift, immediate publication of the book.[528] The imprimatur was therefore granted on December 26,[529] and now Olivieri also made known the reason why the imprimatur was to be granted by Frattini — the Holy Office preferred not to make any appearance whatsoever in the matter, but rather to behave as it did after the Galileo trial: "*cioè di lasciar correre questa opinione senza opporvisi*" (that is, to let this opinion make the rounds without opposing it.)[530]

Now it took a few more days until on January 2 of the new year, 1821, the printing was completed and the *publicetur* ("let it be

524 See the minutes of the meeting on December 14, 1820, in Brandmüller and Greipl, *Copernico, Galilei e la Chiesa*, 398. Candido Maria Frattini (1767–1821) was from Rome and was appointed vice-regent and titular archbishop of Philippi by Pius VII. See Ritzler and Sefrin, *Hierarchia catholica*, vol. 7, 306; Moroni, *Dizionario*, vol. 99 (1860), 179–180. Concerning the office and competencies of the vice-regent, see Niccolò Del Re, *Vicegerente del Vicariato di Roma* (Rome: Istituto di Studi Romani, 1976).

525 See Settele, December 17, 1820, in Maffei, *Giuseppe Settele*, 379–380.

526 Ibid., 379.

527 See Settele, December 21, 1820, in Maffei, *Giuseppe Settele*, 380.

528 See Settele, December 22, 1820, in ibid., 380–381.

529 See Settele, December 26, 1820, in ibid., 382.

530 See Settele, December 27, 1820, in ibid., 383.

published") of the vice-regent was issued on the following day.[531] Settele could finally breathe easy, and he parodied the well-known verse by Virgil: "*tantae molis erat terrestrem volvere massam*" (such great trouble it took to turn the earth's mass around.)[532]

[531] See Settele, January 2–3, 1821, in ibid., 384–385. Frattini reported to Turiozzi on January 10, 1821, that he had executed the order; Brandmüller and Greipl, *Copernico, Galilei e la Chiesa*, 397.

[532] See Settele, January 3, 1821, in Maffei, *Giuseppe Settele*, 385. The original verse is found in the *Aeneid*, bk. 1, verse 33: "*Tantae molis erat Romanam condere gentem*" (Such great trouble it took to found the Roman nation). (*P. Vergilii Maronis Aeneidos* I, ed. R. G. Austin (Oxford, 1971), 2.

Chapter 5
Arguments: Pro and Con

Even more important, however, than the external business surrounding permission to print the second volume of Professor Settele's *Elementi di Ottica e di Astronomia* was — and is — the clash of ideas that determined the historical events. They are now presented in the following pages.

Anfossi — contra

After explaining in nine *Motivi*[533] his "no" to allowing Settele's work to be printed, the Master of the Sacred Palace submitted his reasons for it once again and in expanded form to the pope on October 7, 1820. His discourse, which fills twelve pages in the printed *Positio,*[534] contains Anfossi's arguments with their characteristic features, and

533 "*Motivi per cui il P. Maestro del S. Palazzo Apostolico ha creduto, e crede non doversi permettere al Signor Canonico Settele d'insegnare come Tesi e non come semplice Ipotesi a tenore del decreto del 1620 la mobilità della Terra, e la stabilità del Sole nel centro del Mondo.*" The text is in Brandmüller and Greipl, *Copernico, Galilei e la Chiesa,* 310–317.

534 *Ragioni per cui il P. Maestro del S. Palazzo Apostolico ha creduto e crede che non si può permettere la Stampa del Manoscritto del Signor Canonico Settele, che incomincia: Movendosi la Terra intorno al Sole.* The text is in Brandmüller and Greipl, *Copernico, Galilei e la Chiesa,* 336–349. The date is according to Olivieri's opinion in the dossier; Brandmüller and Greipl, *Copernico, Galilei e la Chiesa,* 353; Maffei, *Giuseppe Settele,* 428.

therefore we can limit ourselves to a more detailed presentation and discussion of them.

Anfossi introduces his *Ragioni* with a reference to his official competence to grant permission to print for all products of the press that are to be published in Rome. Then he emphasizes that he derives his criteria for granting or refusing the imprimatur from Church doctrine, which is based on Scripture, Tradition, the Church Fathers, and magisterial decisions. In doing so, however, he cannot refer to the theories of philosophers and astronomers, which are subject to error. Based on these criteria, however, he comes to the same conclusion regarding Settele as was reached before him by the famous astronomer Boscovich, who wrote that it was not allowed here in Rome (!) to advocate the Copernican theory, since it had once been condemned by ecclesiastical authority.[535] This is rather an obiter dictum, for Anfossi also cites again and again in his *Ragioni*, like a "*ceterum censeo*" (furthermore, I propose), a remark by Boscovich that should be evaluated as a fundamental statement. He goes on to say that the famous astronomer therefore also described the immobility of the earth as divinely revealed and thus as an article of faith. Now Boscovich says, "*Telluris quies ut in sacris litteris revelata omnino admitti debet.*" (The earth's immobility as revealed in the Sacred Scriptures must be admitted entirely.)[536] This formulation, however, is open to interpretation, or indeed, in need of it. Certainly it can mean that the earth's immobility is to be held as a revealed truth. But an equally justifiable

[535] "*Demum inierimus rationem superiore anno indicatam, qua quidquid ii ex Telluris motu derivarunt, quem nobis hic Romae olim a sacra auctoritate damnatum amplecti omnino non licet.*" "Finally, that we entered into the reasoning indicated in an earlier year, whereby we are not permitted at all to embrace here in Rome whatever they had derived from the motion of the Earth, which [theory] was condemned long ago by the sacred authorities." Boscovich, *De Aestu Maris*, 4.

[536] Ibid., 33.

translation—which in context is much more plausible—is "the immobility of the earth, as it is revealed in Sacred Scripture." By that—and this is altogether the genuine sense—Boscovich would have meant the sort of immobility of the earth about which the Bible speaks, and that was not an astronomical-physical standstill of the earth but rather one that was experienced in that way by human beings. Olivieri would later establish this convincingly by exegesis.

Anfossi continues by observing that, in his petition to the pope, Settele himself admits that the teaching about the earth's movement had in fact been condemned. How then can anyone allow the printing of his book? An attempt to weaken the import of the 1616 condemnation by pointing out that only the qualificators spoke them, but not the congregation of cardinals itself, cannot succeed, Anfossi insists. The eleven qualificators were not only eminent theologians; they also acted by order of the congregation and their verdict was confirmed by the pope. The condemnation therefore could not have occurred more authentically and solemnly. In this argument, of course, Anfossi is wrong. No relevant document was ever drawn up in the year 1616 or even later on.[537] Anfossi also interprets incorrectly the decree of the Congregation of the Index dated March 5, 1616, because he does not take into consideration the one dated May 15, 1620, which contained the emendations to be made to the book by Copernicus.[538]

Then he cites the Council of Trent, which declares Church doctrine, in particular the "unanimous consensus of the Church Fathers," as the supreme norm for the interpretation of Sacred Scripture.[539]

537 See GT72 f. = after footnote 85 in Part 1, chap. 2 "History."

538 See Pagano and Luciani, *I Documenti del processo di Galileo Galilei*, 102–103 (1616 Index decree). Aso, St. St. E 5 a/b, copied from the original register 311 r-316 v (1620 Index decree).

539 See Walter Brandmüller, "Die Lehre der Konzilien über die rechte Schriftinterpretation bis zum 1. Vatikanum," *Annuarium Historiae Conciliorum* 19 (1987): 13–61.

Thus it is not up to astronomers and physicists to explain the Bible. Therefore, anyone who like him (Anfossi) or Settele, who has made the Tridentine profession of faith, can never, without violating his oath, advocate a teaching that contradicts the Bible as interpreted by the "*consensus Patrum*." The Copernican theory therefore cannot be taught in any way whatsoever except as a hypothesis. Settele, though, sought excuses and pretexts for doing this nevertheless. He thinks that he can cite in favor of this the fact that Benedict XIV in 1757 did not bother to include in the new Index the general prohibition against books teaching that the earth moves.[540] Nevertheless, Anfossi argues, the pope merely decided not to reprint this decree in the new Index for 1758, but did not thereby imply that it was permissible henceforth to teach the Copernican theory; after all, the individual advocates of heliocentrism remained on the Index by name. Nor can it be otherwise, for the Church can change her discipline but never her doctrine. Besides, it is one thing not to publish a decree again, and another thing to revoke it. An example of this is the bull *In Coena Domini*, which since the reign of Clement XIV has no longer been published, although this does not affect its validity.[541]

As we know, the Holy Office decided to grant permission to print Settele's work on the condition that he would add to the text an explanatory note, which would make clear why the condemnations from the year 1616 were no longer an obstacle.[542] This *Nota* — composed jointly by Settele, Olivieri, Turiozzi, and Grandi, as was mentioned — was now likewise vehemently protested by Anfossi.

He thought that it only made matters worse, since it did not even mention the condemnation of Copernicus's teaching as heretical and

[540] See footnote 48 from Part 1, Chapter 5.

[541] See W. M. Plöchl, in *LThK2* 1 (1957): 32.

[542] See minutes of the meeting on August 16, 1820, in Brandmüller and Greipl, *Copernico, Galilei e la Chiesa*, 300–301.

erroneous in the faith, which Anfossi describes as "*il cardine di tutta la difficoltà*" (the point on which the whole difficulty hinges) — whereby, of course, he labors under a historically inaccurate conception of the vote by the qualificators in 1616. He also opines that to trace the censure from that year back to the nonsensical conclusions that resulted from the fact that Copernicus and Galileo had been unable to refute the objections to their theory, was a *theological* absurdity and an offense against the irrevocability of papal decisions about the faith. This argument, too, went nowhere, because no such decision existed.[543] Nonetheless, he insists on it and emphasizes that such a decision in matters of faith could never be based on changeable philosophical premises, as Settele supposed. This too had not been the case at all in 1616, Anfossi argues; at that time it had been determined that Copernicus contradicted Sacred Scripture and that was the reason for the condemnation. But if this contradiction existed in the year 1616, then it existed also in the year 1820. If a different decision were to be made today nevertheless, this would have the worst possible consequences, for the interpretation of Sacred Scripture and the assistance of the Holy Spirit promised to St. Peter and to his successors, which preserves them from error, always remains the same.

Anfossi continues: Now the *Nota* also maintains that the contradiction observed between the earth's movement and Sacred Scripture was only an apparent one. Yet with that, Settele is claiming no less than that adherence to the literal meaning of Sacred Scripture leads to absurdities, and that therefore one must depart from the literal meaning. But the teaching of St. Thomas is opposed to this, for he clearly says, "*Quod sensui litterali Sacrae Scripturae numquam potest subesse falsum.*" (A falsehood can never underlie the literal sense of

[543] See GT126 = after footnote 39 in Part 1, chap. 3

Sacred Scripture.)[544] But now if the *Nota* says that under the given circumstances we must depart from the literal sense, then Settele is opening up thereby a broad way for individualistic interpretation of the Bible, as is typical of the heretics.

Of course, Anfossi never asked himself the question of whether the *sensus litteralis* of Sacred Scripture has as its content in the first place the immobility of the earth, understood astronomically and physically! His counterpart Olivieri will raise it — and correctly answer it in the negative.[545]

Anfossi's next question reads as follows: Do we not insult God Himself, if we assume that He kept mankind in error so long by always saying that the sun moves — a standpoint that would still be ours, had Copernicus and Canon Settele not arisen? No, the Church founds her doctrinal judgments on Scripture and the Church Fathers, and Torricelli's discoveries changed nothing. The modern reader can only shake his head at such a claim.

Anfossi's argument moves on a somewhat adequate level in a further train of thought concerning the inerrancy of Sacred Scripture. Here he relies on Muratori's work, *De ingeniorum moderatione,*[546] and again he cites Thomas Aquinas. He says that in Sacred Scripture there are many statements referring to astronomy, physics, profane

[544] *Summa Theologiae* I, q. 1, art. 10 ad 3: "For by words things are signified properly and figuratively. Nor is the figure itself, but that which is figured, the literal sense. When Scripture speaks of God's arm, the literal sense is not that God has such a member, but only what is signified by this member, namely operative power. Hence it is plain that nothing false can ever underlie the literal sense of Holy Writ." (Translated by Fathers of the English Dominican Province, 2nd and rev. ed., 1920). If Anfossi had read more than just the last sentence, he surely would have noticed that the entire passage speaks against him, but not against Settele.

[545] See GT201 f. = near footnotes 37-41 in Part 2, Chapter 5 .

[546] See L. A. Muratori, *De ingeniorum moderatione in religionis negotio libri tres* (Frankfurt, 1716), 187–199 (lib. 1, cap. 21).

history, or chronology which certainly have no truths about salvation as their content. However, since they come from inspired authors and thus have become part of Sacred Scripture, it would be sinful to suspect them of error. Therefore such statements should be firmly believed as a component of revelation. As an argument against an imprimatur for Settele, this statement, which is correct in itself, would come into question only if the literal sense of the relevant biblical passages spoke about astronomy and physics — but that is not the case. The passage from Thomas cited by Anfossi in this connection does not refer to what Muratori said. Anfossi's argument becomes completely absurd when he now insists once again that the correct interpretation of Sacred Scripture is not determined by the results of natural scientific research, such as those of Kepler, Menton, and Bradley, but rather by the assistance of the Holy Spirit. Moreover, he continues, astronomers themselves are not unanimous, and even if they did agree regarding the determination of the facts, nevertheless they could still be mistaken about their causes. Besides, these causes were known by the Holy Spirit, and yet He said more than eighty times in Sacred Scripture that the sun revolves while the earth stands still. Any telescopic observation is powerless against this.

To support his standpoint, Anfossi now cites the French apologist Jamin[547] and also Louis Racine,[548] whose statements in

[547] Against the rationalistic optimism of his time, Jamin emphasizes that all creatures of God also have a hidden side that is either difficult for us to know or not accessible at all. See N. Jamin, *Pensieri teologici relativi agli errori de' nostri tempi* (Milan, 1782), 245.

[548] "*Des systèmes savans épargnez-vous les frais. Et ces billans discours, qui n'éclairent jamais: Avouez nous plutôt votre ignorance extrême.*" (Spare yourself the expense of the learned systems, and of these brilliant speeches that never enlighten. Admit to us instead your extreme ignorance.) Brandmüller and Greipl, *Copernico, Galilei e la Chiesa*, 345. From Louis Racine, fils, *La Religion* (Lyon: Rusand, 1832), chant 5, p. 109. The author had

themselves are of course true, but out of place in this context and devoid of any argumentative force. In a further step, Anfossi disputes the remark in the *Nota* that "today" the Copernican system is taught differently in comparison with Galileo's time. He says that now as before it concerns the same thesis: that the earth moves around the sun. Therefore the same condemnation continues in force. Although the *Nota* claims for itself the authority of Cagnoli,[549] nevertheless the much more famous astronomers Tycho Brahe und Boscovich were opposed. Again he refers to Nieuwentijt[550] and Gassendi, who are convinced of the probability of the Copernican system, but do not hold that it is strictly proved.[551] Tiraboschi, also cited by the *Nota*, was once imprudent enough to do so, but nevertheless thought better of it when he was criticized by Fr. Mamachi.[552] Cardinal Gerdil is also unfairly claimed by the *Nota* (and here Anfossi was right for once).[553] Finally Anfossi also calls as his witness the Church historian Alexander Natalis, who relied on Augustine and had demonstrated that the opinions and systems of the

consulted a concordance of the drama and poetry of *Jean* Racine, the poet's father.

549 Cagnoli's *Notizie astronomiche* had been widely circulated. See U. Baldini, in *DBI* 16 (1973): 327.

550 Nieuwentijt personally subscribes to the standpoint of Tycho Brahe. See B. Nieuwentijt, *L'existence de Dieu, demontrée par les merveilles de la nature, en trois parties* (Amsterdam, 1727), 390.

551 Here too the possibility of an increase in knowledge over the course of 150 years does not even occur to Anfossi!

552 About Tiraboschi, see footnote 66 from Part 2, Chapter 1. The remark that he had changed his mind on account of the objections of the then-Master of the Sacred Palace, Fr. Mamachi, testifies to a certain naiveté on the part of Anfossi. An intelligent reader would have noticed the irony with which Tiraboschi replied to Mamachi! See G. Tiraboschi, *Storia della Letteratura Italiana*, 388–389.

553 Gerdil does praise Galileo, but he also remarks that the Holy Office wanted his teaching to be understood only as a hypothesis. See Gerdil *Storia delle Sette de'Filosofi*, 258–259.

philosophers had to prove themselves by Sacred Scripture, and not vice versa.[554] If however Guglielmini and Calandrelli have published their proofs for the earth's movement,[555] they themselves must take care as to how they can reconcile this with the Tridentine profession of faith. Then too, if in the early part of the century even Rousseau's *Du contrat social,*[556] *Cronache dal Paradiso,*[557] Casti's *Novelle galanti,*[558] and other bad books were printed, yet their publication in Rome certainly did not mean that the Church approved of their content. The same is true for those works that believe they can claim the support of the *Nota.*

Olivieri — pro

The objections of the Master of the Sacred Palace had already been invalidated by Olivieri, then still the "*primo Compagno*" (first associate) of the commissioner of the Holy Office, with his *Riflessioni* which bear the date June 10, 1820, and take up fifty-three pages in the printed internal dossier of the dicastery.[559] There is an additional, seventy-page, handwritten treatise about the same question, which however for lack of more precise evidence cannot be classified unambiguously in the proceedings. Although in part it reads thematically the same as the

[554] See Alexander Natalis, *Historia ecclesiastica Veteris Novique Testamenti ab Orbe condito ad A.D. 1600* (Paris, 1730), 1:301; and also Dissertatio XIII "De admirabili statione Solis imperante Josue," in ibid., 2:43–46.

[555] See footnotes 69-70 from Part 2, Chapter 1.

[556] An Italian translation of Rousseau's *Social Contract* had appeared in Rome in 1798.

[557] The work appeared without indication of an author, date, or place of publication. It was included in the appendix to the 1806 Index. See Reusch, *Der Index der verbotenen Bücher*, II/2, 1016.

[558] Giambattista Casti (1724–1803) had published in Rome in 1790 one part of his *Novelle galanti* (complete edition 1804 in Paris). See S. Nigro, in *DBI* 22 (1979): 32. In 1804–1805 his "dirty business" was put on the Index. See Reusch, *Der Index der verbotenen Bücher*, II/2, 1016.

[559] See Brandmüller and Greipl, *Copernico, Galilei e la Chiesa*, 225–287.

previously mentioned work, it nevertheless advocates the interpretation that the decrees from 1616 and later were now as before in force, yet modern astronomy does not contradict them.[560]

Three months later Olivieri wrote a thirty-one-page reply to Anfossi's *Motivi,* in which he refuted them point by point with his characteristic acumen. For all the courtesy of the tone, the objective harshness of the remarks is striking.[561] The commissioner had treated Anfossi's letter to Turiozzi dated August 28, 1820, in a similar manner.[562]

In his responses Olivieri shows his wide reading, his exact and subtle argumentation, but also his rhetorical-polemical skill, which of course his opponent had given him plenty of opportunity to demonstrate.

As mottos for his remarks Olivieri chose statements by two unquestionable authorities: Thomas Aquinas and Augustine. They also reflect the methodological principles that Olivieri intended to follow in his argumentation: "*Multum autem nocet talia quae ad pietatis doctrinam non spectant, vel asserere vel negare quasi pertinentia ad sacram doctrinam.*" (It is very harmful to assert or to deny things which have nothing to do with the teaching of piety as though they did pertain to sacred doctrine.)[563] With that, all theological argumentation in connection with questions of natural science — here astronomy and physics — were dismissed from the start as illegitimate.

The second quotation, taken from Augustine, goes even further, when the bishop of Hippo says, "*Quidquid ipsi de natura rerum veracibus documentis demonstrare potuerint, ostendamus nostris*

[560] Ibid., 184–224.

[561] Ibid., 317–325.

[562] Ibid., 307–310.

[563] "*Responsio ad lectorem Vercellensem de articulis 42 Proemium,*" in *S. Thomae Opera omnia*, ed. R. Busa (Stuttgart-Bad Cannstatt, 1980), 3:640.

litteris [Sacrae Scripturae] non esse contrarium." (Let us show that whatever they were able to demonstrate about the nature of things by genuine proofs is not contrary to our Scriptures.)[564]

Robert Bellarmine had already taken these and similar statements by Augustine as the basis for his interpretation, which he had expressed long before the conflict over Galileo; it had then prompted him with regard to Foscarini and Galileo to demand first the presentation of conclusive proofs for heliocentrism before thinking about an interpretation of the Bible different from the conventional one.[565]

Although in Olivieri's following remarks no strict systematic approach or logical sequence is recognizable and he appears to proceed instead by way of association, nevertheless he does treat — quite thoroughly and with good documentation — all the questions that were or could be asked in that context.

Olivieri's first reference pertains to the source of Anfossi's resistance. It was the work *Summa philosophica* by the Dominican Salvatore Maria Roselli.[566] In it Roselli also speaks about Copernicus and cites decrees that had been issued against his theory. Remarkably, though, he overlooks the fact that the general prohibition of books that teach Copernicanism had been rescinded by Benedict XIV in 1758.

The important Parisian astronomer Lalande traveled to Italy and Rome in the year 1765 and afterward wrote about his stay; Roselli accused him of error.[567] Now, since Anfossi had relied on Roselli's discussion exclusively, the Master of the Sacred Palace had

[564] Augustine, *De Genesi ad litteram libri duodecim,* I, 21, *CSEL* 28/I (1894), 31.

[565] See GT62-64 = from 2 paragraphs before footnote 59 to around footnote 63 of Part 1, Chapter 2.

[566] See Salvatore Maria Roselli, *Summa philosophica ad mentem Angelici Praeceptoris S. Thomae Aquinatis,* 6 vols. (Rome, 1777–1783). Olivieri used the second edition (1785). About Roselli, see P. v. Loë, in *KL2* (1897): 1274.

[567] See Lalande, *Voyage,* 5:48–49.

committed an unforgivable error, as his critic Olivieri reminded him at every step.

Another attack follows. This one has to do with the principles of biblical exegesis — namely, the "consensus of the Church Fathers" cited by Roselli and Anfossi as a norm of interpretation.[568] Now Olivieri turns the tables and asks whether it is conceivable that books teaching heliocentrism could ever have been stricken from the Index if their teaching in fact had contradicted the "*consensus Patrum.*" Olivieri anticipates the expected objection that they were in fact forbidden in 1616 and after, and so how could anyone have permitted them in 1758,[569] by pointing out that their teaching then was still incomplete, but now it has been refined and corrected. This important idea is central to Olivieri's argument.

In the following discussion, which now really advances to the theological center of the controversy, Olivieri proves to be a specialist in the field of Old Testament hermeneutics and exegesis, since he sets out to invalidate the Old Testament arguments against the earth's movement that were adduced by Roselli and by Anfossi after him.

First he points out that Sacred Scripture has visible phenomena in mind and makes use of everyday colloquial language whenever it speaks about the rising and the setting of the sun. This fact allows us to explore the real astronomical-physical causes for what can be seen in the heavens, without thereby offending against Sacred Scripture. The same is generally true also when it is a question of patristic texts.

Since such astronomical questions are by no means part of divine revelation, the discourses of the Church Fathers in this

568 See Brandmüller and Greipl, *Copernico, Galilei e la Chiesa*, 339.

569 See ibid., 340.

regard had just as much weight as the proofs they offered, whether or not they taught unanimously. For this methodological presupposition Olivieri could cite Melchior Cano, the classic writer on theological methodology.[570] He continues by noting that anyone who nonetheless insists on a contradiction between Scripture and the Church Fathers, as Copernicus did, cannot ever have studied Scripture or the Fathers seriously. Otherwise, how could he have missed the poetical character of a series of passages and the use of rhetorical stylistic effects, or furthermore the fact that neither Scripture nor the Church Fathers intended to impart astronomical teaching about the construction of the universe? As an illustration of this the commissioner of the Holy Office cites several passages from the work of the exegete Calmet,[571] in the Italian translation in which they were added to the 1744 Paduan edition of Galileo's works.[572]

From this, he says, it follows clearly enough that it is the intention of Sacred Scripture, by its depiction of celestial phenomena, to direct our attention to God, the Creator and upholder of the universe.

After these methodological preliminaries, Olivieri proceeds to discuss the individual pertinent Bible passages.

[570] See Melchior Cano, *De locis theologicis libri duodecim*, 7 vols. (Rome, 1890; original edition Salamanca, 1563), 3.

[571] Augustin Calmet, O.S.B. (1672–1757) was from Lothringen, abbot of Senones, and one of the most famous exegetes of the eighteenth century. His magnum opus, *La sainte Bible en latin et en français avec un commentaire littéral et critique* (Paris, 1707–1716) ran to twenty-three quarto volumes. It was reprinted many times. In the present context, Olivieri cited his *Dissertations, qui peuvent servir de prolégomènes de l'Écriture sainte* (Paris, 1720), which were translated into English and Dutch. See P. Volk, in *LThK2* 2 (1958): 886.

[572] See *Le Opere di G. Galilei* 4:1–20.

First he examines Ecclesiastes (Qoheleth) 1:4–6: "The earth remains [stands] forever." It is easy for him to show that the topic here is by no means astronomy, not even the universe, because the preceding line, "A generation goes, and a generation comes," makes the purpose of the statement clear. It is about the contrast between transient mankind and the earth, between what is passing away and what is the permanent, invariable arena of human development. When it says next, "The sun rises and the sun goes down, and hurries to the place where it rises" (Eccles. 1:5; NRSVCE), the question that arises is not whether this corresponds to appearances, but rather to what physical causes the observations can be traced. But this same question plays no role in the scriptural context. Its real message has to do with the constant recurrence of natural phenomena, their "everlasting circulation." This is clear from the context offered by the following verses.

It is even easier for the exegete to upset the "probative weight" of two other Bible passages, namely 1 Chronicles (Paralipomenon) 16:30 and Psalm 93:1. Again these speak about the earth standing still, and again the context shows that it is impossible to see them as statements about astronomy or physics. Rather, Olivieri says, these verses likewise speak about the earth as the habitat of man, which in fact is not shaken, as the psalm says. This would be possible only assuming the "*mobilità scompigliatrice e scompaginatrice*" (disordering and breaking up movement) in the theologians' objection,[573] which however is not correct.

And now a new argument: that the decree from 1620 permitted the assumption of the earth's movement as a hypothesis. This would not have been possible if the earth's movement *in itself*

[573] That is, if the earth's movement has as a consequence that everything on the earth's surface is thrown around in a furious swirl.

contradicted Sacred Scripture and not only in connection with its "*scompiglio terrestre*" (earthly disarrangement), as was assumed then. Something like that could not have been admitted even as a hypothesis. This, however, goes to prove that a movement of the earth, apart from these accompanying phenomena of swirling and confusion — everything that takes place on the earth's surface — in no way contradicts Sacred Scripture. This same separation of the earth's movement and "*scompiglio*" was made possible by the later discovery of the weight of the atmosphere, which rules out the latter absurd effect.

Then Olivieri discusses the oft-cited miracle of the sun in Joshua 10:12–15: " Sun, stand still at Gibeon/ Moon, in the valley of Aijalon!/ The sun stood still, the moon stayed,/ while the nation took vengeance on its foes." Here Isaiah 38:8 should be kept in mind, too: "See, I will make the shadow cast by the sun on the stairway to the terrace of Ahaz go back the ten steps it has advanced. So the sun came back the ten steps it had advanced." Now the exegete poses the question, whether we may conclude from the facticity of the miracles reported in these passages that the physical circumstances underlying them are also revealed. But the sacred texts are silent about this! Consequently it is the business of the astronomers to explore these phenomena.

It was easy to refute a final thought by Fr. Roselli, who tried to conclude from the wording of the creation account (Gen 1:14–19) that the earth is the foundation and center of the universe. All Olivieri needed to do was to cite the text of Genesis 1:1: "In the beginning God created the heavens and the earth." (RSVCE)

Therefore, first the heavens and then the earth. Presumably Olivieri was not entirely at ease with this exegesis, in which he descended to the level of his opponent, for he also cites Thomas

Aquinas[574] and the astronomer Lalande.[575] Thomas formulated a hermeneutic principle that should be applied here, when he pointed out that Moses spoke about visible phenomena in order to make himself understood by his uneducated people.[576] Neither he nor any other writer of Sacred Scripture intended to teach an astronomical system. Notwithstanding the authority of Sacred Scripture, this is a matter for the astronomers. This concluded the refutation of the biblical objections to Copernicus, whereby without the aid of other "scriptural senses" Olivieri — being a philologist — focused exclusively on the literal sense and thus complied with the demand that had already been made by Galileo's contemporary opponents.

Next he took up the task of invalidating Anfossi's thesis that the unanimous teaching of the Church Fathers, the "*consensus Patrum*," also forbids the acceptance of the heliocentric system; Anfossi had relied on Alexander Natalis to support this thesis. An unbiased reading of the cited text, however, led Olivieri to find nothing but a statement that the Church Fathers being quoted emphasized the

574 See *Summa theologiae*, I–I, q. 70, art. 1 ad 1 (not ad 2, as Olivieri writes). This deals with the question of whether the lights of heaven, in other words the sun and moon, were created on the fourth day of the Hexaemeron (six-day Creation) or "*ante omnem diem*" (before any day, i.e., before time began). *S. Thomae Aquinatis Opera omnia, ed. iussu impensaque Leonis XIII. P.M.* (Rome, 1889), 5:177. Thomas rejects the question itself, citing Augustine: "For he [Augustine] does not suppose a succession of time in these works" (ibid.). The fact that sun and moon are not mentioned until the fourth day (!) has no significance: Chrysostom says that it is only a matter of showing the people that they are not deities; in other words, that they were created and therefore did not exist from the beginning (ibid., 177 –178).

575 See Lalande, *Astronomie* I, 536–541, esp. 539.

576 See *Summa theologiae* I–I, q. 70, art. 1, ad 3: "Yet our senses perceive the movement of the luminaries and not that of the spheres. But Moses describes what is obvious to sense, out of condescension to popular ignorance, as we have already said."

facticity of the miracle reported in the book of Joshua, and that Alexander Natalis was arguing thereby against the exegesis of Rabbi Levi and of Maimonides, on whom Grotius too relied. Alexander Natalis is therefore defending the facticity of the miracle, but he does not say a single word about its physical causes.[577]

Olivieri verified the same thing when analyzing the patristic passages that Riccioli had cited:[578] they did not address the subject dealt with here. We must recognize clearly, he continues, that many Church Fathers had no accurate notion at all of the spherical form of the earth, but simply went along with the "*concetti non rettificati*" (uncorrected concepts) of their contemporaries. The great Montfaucon therefore says in the introduction to his edition of *Cosmas Indicopleustes* that the Church Fathers in their statements concerning astronomy had altogether different notions, depending on whether they were well informed or had instead made fun of the experts (although they themselves were uninformed).[579]

Although that is also true, we can nevertheless find in the writings of the most enlightened Fathers excellent instructions for observing things correctly. For example, Basil, after several discussions about the structure of the cosmos, exhorts his readers to admire the wisdom of its Architect, just as John Damascene sets aside all theories concerning this and says that whatever the case may be, all this certainly came about at God's command.

577 See footnote 22.

578 See Riccioli, *Almagestum novum*, 2:481–486 (lib. 9, sect. 4, cap. 37).

579 Olivieri had available a copy of the *Christiana Topographia sive Christianorum opinio de mundo* by Cosmas Indicopleustes in the edition by B. de Montfaucon, *Collectio nova Patrum et Scriptorum Graecorum II* (Paris, 1706), 113–345 (preface: iv–vi). It is most easily available today in *PG* 88 (1860), 51-476. See H. G. Beck, *Kirche und Theologische im Byzantinischen Reich* (Munich, 1959), 416 f.

After this Olivieri cites only Augustine, of whose many works *The Literal Interpretation of Genesis* was of course relevant here: he already dared to understand the six days of the work of Creation in a way that was not literal.

It is not so essential, therefore, but rather rhetorical when Olivieri then points out that Anfossi contradicts himself, since he is convinced that the earth is motionless yet several times calls it "our planet." At the same time, though, he says that in the miracle related in Joshua, the sun, or (as others thought) the earth had stopped its course (!). If he demands furthermore that we must adhere in this question to the literal sense of the Bible, until the Church decides otherwise, then he thereby does admit the possibility of an explanation other than his own. Flagrant inconsistencies! However, since Anfossi in this way admits that the earth's movement is still possible, then the judgment of the Inquisition in 1632 now affects him, too: "*Qui pariter est gravissimus error, cum nullo modo probabilis esse possit opinio, quae iam declarata, ac definita fuerit contraria Sacrae Scripturae.*" (Which is likewise a very serious error, since an opinion which has already been declared and defined as contrary to Sacred Scripture can in no way be probable.) But with such logical gymnastics, which Olivieri himself probably did not take altogether seriously, he was in any case able to exasperate Anfossi. There was little to be proved in that way.

But now he starts a more serious interpretation of the 1633 decree. After thoroughly summarizing the course of events, he quotes the important figure Tiraboschi, who calls the judgment against Galileo a "*troppo rigorosa censura*" (overly rigorous censure) and emphasizes that it was passed by a tribunal to which even the most zealous Catholic had never attributed infallibility.[580] This is a

[580] See Tiraboschi, *Storia della Letteratura Italiana*, 228.

sign of Divine Providence, which did not allow a solemn judgment of faith to be made in this matter at a time when most theologians were convinced that Copernicus taught contrary to Scripture. Olivieri agrees and therefore rightly contradicts Roselli's claim that the sentences issued against Copernicus and Galilei are irrevocable.[581] But all the same, for this reason he cannot regard the sentences as wrong. He recognizes with increasing certainty that we are dealing here not with a simple matter, as it may seem at first glance, but rather with a very complex problem. We must therefore carefully untangle and keep separate the individual ideas that are intertwined. In doing so we will find that the ideas that were condemned deserved it, while others were not affected by the condemnation. And the latter are precisely the ones that were taken up by modern astronomers, when they spoke about the movement of the earth and the immobility of the sun.

In order to carry out this analysis, Olivieri turns first to the 1616 decree of the Congregation of the Index. This one articulated two *qualificationes* with regard to the teachings of Copernicus. The one reads "*falsa*," and the other "*contraria alla Sacra Scrittura*." What these two qualifications referred to in particular is made clear by the vote of the experts, which is summarized in the aforesaid Index decree and is considerably defused. The text by the experts says first that the heliocentric theory is "*absurda et falsa in philosophia*," and then—with reference to the earth's movement—this teaching is "*formaliter haeretica, quia est expresse contraria Sacrae Scripturae*" (formally heretical, because it is explicitly contrary to Sacred Scripture). Regarding the immobility of the sun, the text continues with, "*et theologice considerata ad minus erronea in fide*" (and considered theologically, at least erroneous in faith). This makes it clear that the term *falsa* refers to the

[581] See Brandmüller and Greipl, *Copernico, Galilei e la Chiesa*, 228.

philosophical absurdities that are not the object of theological knowledge—that is, the kind derived from revelation—but rather of the philosophical knowledge that can be gained in the light of natural reason. The qualification "*contraria alla Sacra Scrittura*" is another matter. This is a theological statement, which results when we apply the standard of Sacred Scripture to the theory under discussion. Such a judgment, however, presupposes the factual demonstration of the philosophical falsehood. Only if someone were to give a philosophically absurd meaning to the statements of Sacred Scripture would this be a contradiction of Sacred Scripture. Therefore it is a question of determining the philosophical falsehoods and absurdities that the Congregation of the Index and the Holy Office had in mind in Galileo's day, and then these are the things that were in contradiction to Sacred Scripture, as the sentence says. However, if these falsehoods and absurdities can no longer be found in the teaching of modern astronomers, then their teaching is no longer the teaching that was once condemned, and also it no longer contradicts Sacred Scripture.

Which was the case, therefore, Olivierii now asks, with the falsehoods and absurdities that were objectionable then? Since the files of the Galileo trial did not return from Paris,[582] we must instead have recourse to the literature from Galileo's time. First of all, however, it must be stated that the stumbling block must not be sought in the movements of the heavenly bodies, for it was quite permissible to cite the Copernican system in the form of a hypothesis to explain them. Francis Bacon, Baron of Verulam, wrote that Copernicus cannot be refuted by astronomical principles; this is possible only through the correct application of the principles of natural philosophy.[583] Olivieri investigates these principles in the following discussion. He finds them in the

582 This first happened in 1843. See D'Addio, in *DBI* 6 (1964): 2–3.

583 See Francis Bacon, *De dignitate et augmentis scientiarum* (Frankfurt, 1665), 98 (lib. 1, ap. 1).

works of the contemporary opponents of Galileo. Among them Riccioli seems to him to be the most important. In his *Almagestum novum*, Riccioli compiled most of the arguments then being formulated against Copernicus. The weightiest ones were derived from the absurdities that would necessarily result (given the presuppositions then) from the assumption that the earth moves. If the earth revolved furiously and was swept along rapidly from west to east, then what would happen to the heavy or light bodies on its surface? What about the air, which such a movement would leave behind? The unavoidable result of such a movement of the earth would be a tremendous collision of all the bodies found on the earth's surface. Moreover, a ceaseless, shrill, piping noise would be audible; no living thing could stand on its feet; trees and houses would lie flat on the ground. No bird could fly toward the east, no ship could stay on course, the clouds would always drift back toward the west...

Neither Copernicus nor Galileo could rule out all these absurdities. Torricelli, however, after the death of his teacher, Galileo, successfully demonstrated that air has weight, and that finding, Olivieri insists, now disposes of all those objections. Even so, Olivieri notes, for a rather long time people did not recognize the full import of this discovery and its significance for the heliocentric system, as proved by the opinions against Copernicus that could still be heard in Olivieri's time.

But now Olivieri sets out to examine Fr. Roselli's anti-Copernican arguments in the light of these facts. It is a long, tiring argument, which is undertaken chiefly with the methods of astronomy. The evidence gained in this way, he then says, had already convinced the famous Nikolaus Cusanus that the earth moved, almost a century before Copernicus.[584] The result of his analysis: the arguments that

[584] See *De docta ignorantia*, 11–12, in Nikolaus von Kues, *De docta ignorantia/ Die belehrte Unwissenheit*, vol. 2, ed. P. Wilpert and H. G. Senger, Schriften des Nikolaus von Kues XV b (Hamburg, 1977).

the learned Fathers Roselli, Pini,[585] and Schettini[586] cite against him turn out to be, in the light of the new discoveries, a proof of the Copernican theory. Moreover, an author like L. L. Mercier[587] should not be taken seriously.

Now, though, Olivieri must embark on a discussion about a passage from St. Thomas, which is likewise cited by Roselli. He means the famous section from the commentary on the *De caelo et mundo* by Aristotle, which from the beginning has often been quoted in the present context. The point here is that ancient astronomers had tried respectively in different ways to explain the irregularities in the orbits of the planets. Thomas comments on this as follows: although it may be that these attempted explanations seem plausible, nevertheless we must not say that they are in fact true, for it is also quite possible that there is another accurate explanation that is not yet known.[588] Obviously, Olivieri says, Thomas was convinced that such an explanation could yet be found. In fact the orbits of the planets were also what prompted Copernicus to develop his system. Thomas, on the other hand, already pointed out the way for Cusanus and Copernicus as early as the thirteenth century. This proves that Roselli strayed from the teaching of his "master" Thomas on this question.

With some satisfaction, Olivieri observes the same thing regarding another text by Thomas. The passage in question is Aquinas's answer to the forty-two questions of John of Vercelli. The latter asked Thomas to explain whether the passage Ecclesiastes (Qoheleth) 1:6 — "*in circuitu pergit spiritus*" (the spirit . . . returneth to his circuits),

585 See D. Pini, *Esame del Newtoniano sistema intorno al moto della terra* I-III (Como, 1802).

586 See A. Schettini, *Elementa physicae*, vol. 2 (Rome, 1777), 118, 121.

587 See L. L. Mercier, *De l'impossibilité du système Astronomique de Copernic et de Newton* (Paris, 1806).

588 See *Commentaria in libros Aristotelis de caelo et mundo*, lib. II, lect. XVII, in *S. Thomae Aquinatis Opera omnia* 3:186–189.

could be interpreted to mean the following: the spirit (i.e., the angel) moves in heaven in a circle and thereby moves heaven itself in a circle. Aquinas replies: there is no reason not to interpret the passage in this way; after all, Augustine says that there are many ways to interpret the words of Scripture, so that those who are puffed up by worldly knowledge must refrain from ridiculing it.[589] Thus Olivieri once again turned the tables, since this principle could and indeed must refer also to the well-known biblical objections to Copernicus.

And now the summation of what Olivieri presented thus far: Surely it has been proved satisfactorily now that the arguments once opposing the Copernican system have meanwhile become untenable. However, the ecclesiastical condemnations of his time were based on them, and thus today they are without foundation. Anyone who still makes these arguments only shows that he does not understand what he is talking about — or that he is speaking against his better knowledge, as we can demonstrate to Roselli on account of his interpretation of the 1616 decree and of the 1758 Index, which very wisely no longer listed works that teach heliocentrism among the forbidden books.

But how can we explain the fact, Olivieri goes on to ask, that notwithstanding this development several relevant books remained by name on the Index? The answer is that these books — specifically by Copernicus, Kepler, Foscarini, Zúñiga, and Galileo's *Dialogo* — represented the state of research at the time when they were

[589] See *Responsio ad lectorem Vercellensem de articulis 42*, art. 18: "Decimus octavus articulus est, an illud Eccle. 1,6 'In circuitu pergit spiritus,' sic possit exponi: Spiritus, puta angelicus, pergit in circuitu caeli, et pergendo caelum movere facit in circuitu. Non video quare haec expositio sustineri non possit; praesertim cum dicat Augustinus in I super Gen. ad litteram cap. XX, quod ideo multis exitibus verba Scripturae exponuntur, ut se ab irrisione cohibeant litteris saecularibus inflati." *S. Thomae Aquinatis Opera omnia*, 3:641.

produced, and back then the Copernican system still labored under the oft-cited absurdities.

But what are we to think about the recantation that the great Galileo had to make? On this topic Olivieri says that it is certain that Galileo was able to act in that way without coming into conflict with his conscience, and that an astronomer today could likewise without difficulty make the same recantation, as far as its theological content is concerned. For what did Galileo really regret and abjure in his recantation? First, that he disregarded the prohibition issued in 1616 against teaching heliocentrism, which was true. Then he had to confess that in precisely that way he had incurred the suspicion of heresy. This was true, also. Thematically Galileo retracted two theses: first, that the sun stands in the center of the universe and does not move, and second, that the earth is not the center and it does move. Interestingly enough, this formulation was notably different from the one recommended by the qualificators of the Holy Office, who in reference to the first thesis had spoken explicitly about a local motion — that is, about a movement of the sun from east to west — while their formulation of the second thesis had read: the earth is neither the center of the universe nor immovable, but rather moves, indeed daily.

In order to ascertain what Galileo positively acknowledged thereby, we must look at the antithesis of what was retracted, deducting the absurdities that were perceived then, which Galileo too would have rejected, had he known them. This now results in a meaning which Galileo could hold without any dishonesty. This antithesis reads as follows: The sun is not located in the center of the universe and does move. But that is entirely in keeping with today's astronomical knowledge, according to which the sun is one of the two foci of the elliptical orbit described by the planets. It moves not only around its own axis, but also on an orbit together with its planets and their moons, heading toward the constellation Hercules. Even though Galiileo could not yet have known this, he must have recognized nevertheless that the sun could not be considered the central point in relation to the fixed stars, since they did not move around it. Galileo also knew that earthly bodies fall toward the center of the earth and not toward the sun. Therefore he was able to retract the second thesis, too. All that he had to profess — and he could — was that the sun is not immobile. All the rest was not included in the formula of recantation.

As far as the earth was concerned, the positive antithesis had to read as follows: the earth is the center and does not move. But who does not understand clearly that the earth is the central point for all bodies that surround it and fall toward it? The center for water and air? Also with regard to heavenly phenomena, for which the earth, from our vantage point, really is the center? Significantly, Galileo did not have to acknowledge the earth as the center of the universe, but only as *a* center!

When Olivieri wrote this down, he was certainly not aware that with this view he was not far from the starting point of the theory of relativity! As far as the earth's movement is concerned, he continued, the phrase added by the qualificators, "*motu etiam diurno*" (also by a daily motion), was not included in the recantation formula. Even apart from this, can't we speak, really, about a sort of immobility of the earth? Isn't it motionless with regard to all that comes to be and passes away on its surface? Doesn't it remain the unshakable foundation for all that happens on it? Is it not motionless with regard to falling bodies? Can't we say that it does not move, since as a whole it remains stable despite all its movement? This is true also with regard to the air. Finally we can call the earth motionless with regard to its relation to the other heavenly bodies, too. This relation has an unchangeable regularity, without any disorder appearing in the heavens because of a cessation of the movement. Galileo could swear to all this with a clear conscience. The formulations of the theses therefore allowed altogether favorable interpretations, according to which the old judgments against modern astronomy no longer applied.

A further step in the argument leads Olivieri to discuss the question of how the preceding can be reconciled with the immutability of Church teaching, or, to put it the other way around, the question of the development of doctrine. First he emphasizes the immutability of the "deposit of faith," which is represented in an increasingly explicit way by the doctrinal proclamations of popes and councils. Applied to the present case, this means that we cannot refer to the old Fathers of the Church in order to judge it, because no trace of an explicit Church teaching about this subject can be found yet in their writings.

Although the ancient philosophers down to Ptolemy had reflected on the questions of astronomical cosmology and written about them, no indication can be discovered that the same was true of the Church Fathers. Moreover, Lalande points out that in the Roman Empire the natural sciences were neglected anyway. They had found their home instead in Egypt, especially in Alexandria.[590] As proof Olivieri cites a letter by Pope Leo the Great, in which he asks Emperor Marcian to make inquiries with Egyptian and other specialists in order to solve a calendar problem concerning the calculation of the date of Easter.[591] Then Olivieri mentions the controversy about the antipodes at the time of Pope Zachary, who condemned the teaching that a second world and a second mankind existed under the earth, which shows that he had no idea

[590] See Lalande, *Astronomie*, 1:146–160 (the importance of Alexandria for astronomy), and 155 (the Romans' low opinion of astronomy).

[591] See Leo the Great to Marcian, June 11, 453, in *Acta Conciliorum Oecumenicorum*, ed. E. Schwartz, II/4 (Berlin-Leipzig, 1934), 75–76.

about the spherical shape of the earth. The condemnation of Virgilius of Salzburg that he pronounced therefore has been dropped, without it necessarily being described as false in itself.[592]

In later history he, Olivieri, finds no other relevant facts until we reach Nikolaus Cusanus and Copernicus, whose book was recognized for seventy years before it was suddenly censored. If Copernicus had in fact erred in faith, would it have been necessary to wait so long for the *grido della fede* (outcry of faith)? [593] Further remarks about Copernicus follow, whereby Olivieri pointed out in particular the importance of the astronomers of Frauenburg for the calendar reform in the year 1582. From the start this reform enjoyed the high esteem of popes and cardinals. This means that the Church, the Curia, and the popes had had no objection to the teaching about the earth's movement for two hundred years, from Nikolaus Cusanus to Galileo. And even then there had been only a conditional censure, whereby the value of the Copernican system was still acknowledged, provided that it was presented in the form of a hypothesis.

Even when Galileo was condemned on account of his disobedience to the 1616 prohibition, in his recantation he did not have to acknowledge anything that truly would have contradicted the Copernican system. Nor can we discern any indication that the Holy See condemned the two movements of the earth in their true, proper meaning. This is even more true for the time after the Galileo judgment. Olivieri emphatically underscores the fact that in all the years since 1633 he could discover not one single measure of the Congregation of the Index or of the Holy Office against one of the numerous

592 See F. Lakner, in *LThK2* 1 (1957): 658.

593 An allusion to the title of an anonymous publication by the Master of the Sacred Palace: [F. Anfossi], *Il grido della fede contro la metafisica sublimiore de Deo trino e uno* (Monaco, Rome, 1820). The work runs to thirty-one pages.

books teaching heliocentrism that had appeared since, or against their authors.

Then Newton made decisive progress with his discovery of gravitation, which modified the Copernican system in such a way that we could call Copernicus a Newtonian. The fact that Newton was never condemned, although this absolutely ought to have happened if the old sentences had applied to his teaching, shows that they did not fall under the judgments passed in 1616 and 1633.

Besides, Olivieri now cites as testimony in favor of the actual tolerance of heliocentrism a good number of facts that we have already reported elsewhere. Of course, when he claims that he found in the editions of the Index between 1664 and 1752 no general prohibition of books that taught the movement of the earth and the immobility of the sun, then we must nevertheless accuse him of a rather superficial reading. Obviously the prohibition is found in every Roman edition of the Index, until in 1758 for the first time it no longer appeared. Therefore the principle that he now cites, through valid in itself, could not be convincing: namely, that human laws lose their validity through forgetfulness or non-observance. He also opines that we cannot accuse of indolence the many excellent popes who remained silent on the Copernican question in the last two centuries. This argument may pass, for none of these popes urged in any case whatsoever the general prohibition of heliocentric books.

Thus — and now Olivieri comes to the conclusion — Anfossi strayed far from the truth. For him there is not one single reason to withhold the imprimatur from Settele's work, since books along the same lines that have been published in the last thirty years met with no objections whatsoever and received the imprimatur. Notwithstanding all this, if Anfossi relies on Roselli, whose philosophical and theological wretchedness was made amply clear, and now as before opposes the printing of Settele's *Astronomy,* he only heaps shame on the Holy See, indeed, on the Catholic religion in general.

If we disregard a few subtleties of his reasoning that are not immediately comprehensible, we must acknowledge that Olivieri framed very precisely the question that was appropriate here. Furthermore he manifested great erudition and acumen in developing his argumentation. Thus he probably succeeded in showing convincingly that the Holy See censured heliocentrism in the year 1616 for reasons that were just as good as the ones it applied in 1820 in permitting it, without incurring the accusation of changing its position. Through this proof and above all by the manner in which he conducted it, Olivieri was a credit to the Holy Office.

Chapter 6
Causa Finita

After all these battles, the copies of Settele's *Elementi di Ottica e di Astronomia*, volume 2, fresh off the press, could therefore be delivered by the de Romanis printing house on January 10, 1821.[594] Overjoyed and with considerable pride — Settele reveals in his diary — he now brought his book to friends and patrons and in exchange received many compliments. His great hour struck, however, on January 24, when he was received by Pius VII in a private audience.[595] The pope took the book and placed it on a chair; then he began a conversation that can scarcely be described as serious. It confused Settele, who plainly did not understand the pope's humor and irony. But what else did the pope mean when once again he expressed doubts about the demonstrability of heliocentrism and said, grinning, that he read in a book by Voltaire — which he had better left unopened — that Friedrich II of Prussia had once told d'Alembert that even the natives in Madagascar cited proofs for the movement of the sun![596]

What disappointed Settele most was that the pope bestowed no medal on him, which he had hoped and longed for. Nor did he

[594] See Settele, January 10, 1821, in Maffei, *Giuseppe Settele*, 393.

[595] See Settele, January 24, 1821, in ibid., 401.

[596] Pius VII had already referred to this remark by the Prussian king in speaking with Cristaldi during the previous year. Settele, February 20, 1820, in ibid., 295.

receive a benefice — another wish that he had often confided to his diary.

But Settele decided to dedicate his work not only to the pope but also to Emperor Franz I of Austria, after having had the privilege of giving the latter a copy of the first volume while he was staying in Rome.[597] Therefore he turned to the Austrian ambassador, Count Anton Apponyi, with the request that he forward the book to the emperor, which he did on January 31, 1821.[598] When Settele received no reply, he spoke to the ambassador once again in September,[599] and after a long time he received a medal from the emperor, of which he was very proud.[600] Although the great fame that he dreamed of was not his lot,[601] he certainly did not lack recognition in the educated and scientifically interested circles in Rome. Besides the Lincean Academy, one of their centers since the days of Christina of Sweden had been the Accademia dell'Arcadia.[602] As we saw earlier, after the end of the Napoleonic troubles, Prince Pietro Odescalchi[603] had

597 This is clear from the accompanying letter by the Austrian ambassador, Count Apponyi, dated January 31, 1821, to Prince Metternich (HHStA Wien, Staatskanzlei Rom, Administrative Berichte 2 fol. 51).

598 This copy, splendidly bound in calfskin and decorated with gilt cornices and gilt edges, is found in the Austrian National Library in Vienna under the call number 253.939-B Fid.

599 Enclosed in the copy mentioned in footnote 5 is Settele's original letter to Count Apponyi, dated September 14, 1821, in which he requests an imperial acknowledgment of receipt for his book. Apponyi then wrote about this matter to Metternich on September 17, 1821, whereby he cited an intervention of the Dutch ambassador de Reinhold in favor of Settele (HHStA Wien, Staatskanzlei Rom, Administrative Berichte 2 fol. 190).

600 See Settele, July 26, 1822, in Maffei, *Giuseppe Settele*, 411.

601 "In short, my affair should cause a furor in Germany, and I have been immortalized, I have earned a place in the History of Astronomy" (translated from Italian). Settele, April 27, 1821, in Maffei, *Giuseppe Settele*, 407–408.

602 See Biblioteca Apostolica Vaticana, *Tre secoli di storia dell'Arcadia.*

603 Pietro Odescalchi, (February 1, 1789–April 15, 1856). Among his own scientific works we should emphasize a translation of the text of Cicero's

breathed new life into it by founding the *Giornale Arcadico de Scienze* (*Arcadian Journal*) in 1819. Although by education and inclination he himself was an ancient philologist and archeologist, as the publisher of the *Giornale Arcadico* he was very concerned also about reporting new results of research from the field of the natural sciences in the periodical, which appeared monthly. So it is not surprising that the editors also intended to publish in the journal an "*estratto*" — a thematic summary — of Settele's book.[604] They had already printed a thorough and favorable review of volume 1, on optics.[605] Dr. de Crollis,[606] a physician, then submitted in April 1822 a "mediocre" manuscript[607] (Settele's term), which however was to become immediately the occasion for a comedy that must be described as beyond absurd: the Master of the Sacred Palace, Fr. Anfossi, refused the imprimatur for the excerpt from a book which itself had appeared a year earlier with ecclesiastical permission to print from the highest authorities.[608]

De re publica, which had been discovered shortly before by Angelo Mai. He became president of the Pontificia Accademia Archeologica, the Pontificia Accademia dei nuovi Lincei, and the Collegio filologico dell'Università Romana. Since the reign of Leo XII he held political and administrative offices; appointed to the Consulta di Stato by Pius IX; 1849, president of the Municipal Commission. See P. Biolchini, *Notizie biografiche del Principe Don Pietro Odescalchi* (Rome, 1856); Stefano Rossi, "Elogio del Principe Don Pietro Odescalchi, già direttore di questo giornale," *Giornale Arcadico* 118 (May–June 1857): 3–38; Moroni. *Dizionario*, vol. 20 (1843), 9; vol. 48 (1848), 268.

604 See Settele, July 27, 1821, in Maffei, *Giuseppe Settele*, 408–409.

605 During its first year of publication, in the third issue. *Giornale Arcadico* 1, no. 3 (1819): 216–225.

606 About Domenico de Crollis (1786–1862), see D. Bomba, *Necrologia dell Dott. Domenico De Crollis* (Genoa, 1863): A. Monti, *Della vita del prof. Domenico De Crollis* (Rome, 1863); Moroni, *Dizionario*, vol. 85 (1857), 184–185.

607 See Settele, April 12, 1822, in Maffei, *Giuseppe Settele*, 409.

608 See the entries by Settele from May 5 to November 24, 1822, in ibid., 410–414.

After all the intra-curial trench warfare that has been described at length, in which he had long since been defeated, Anfossi grotesquely continued to try the patience of the author, the Holy Office together with the Congregation of the Index, indeed the cardinal secretary of state and the pope, and now also Prince Odescalchi, by refusing the imprimatur for the manuscript by de Crollis, whereupon the editors of the *Giornale Arcadico* turned to the Secretariat of State, which proved to be very well-disposed to their petition, yet wanted to treat Fr. Anfossi as considerately as possible. He, however, declared that the imprimatur for the *estratto* should be given by the one who issued it for the book; he himself wanted nothing to do with it.[609]

On July 26 one member of the editorial board of the *Giornale Arcadico*, Dr. Folchi, reported that Cardinal della Somaglia had given permission to print; Fr. Anfossi, however, was very annoyed at this, and therefore was delaying the ecclesiastical censor's review of the new issue of the *Giornale Arcadico*.[610]

Therefore Prince Odescalchi turned to the Holy Office, which decided at its *Feria Quarta* meeting on August 21, 1822, that the vice-regent should issue the imprimatur for the excerpt cited by de Crollis as he had already done for Settele's book.[611] Thereupon Anfossi issued an instruction to all the printers in Rome in which he forbade them to print manuscripts that did not bear his imprimatur.[612] He intended by this move to prevent the printing of the *estratto*, but the Holy Office responded to it with a decree dated September 11, 1822, in which the cardinals forbade the present Master of the Sacred Palace and his successors to withhold the

[609] See Settele, May 5 or June 15, 1822, in ibid., 410–411.

[610] See Settele, July 26, 1822, in ibid., 411.

[611] See Olivieri's expert opinion, late August 1822, in Brandmüller and Greipl, *Copernico, Galilei e la Chiesa*, 412–426.

[612] See Settele, November 7, 1822, in Maffei, *Giuseppe* Settele, 413–414.

imprimatur from books that taught the movement of the earth and the immobility of the sun. Those who disobeyed were threatened with punishments.[613] However, this did not impress Anfossi in the least. On September 25 he once again issued a negative reply to Prince Odescalchi,[614] and the following day, when the latter turned for help to the assessor of the Holy Office, Msgr. Turiozzi, the congregation instructed the vice-regent to grant his permission to print.[615] This was done immediately. Now, though, the printer was the one who hesitated, because he did not want to come into conflict with Fr. Anfossi, as his business was dependent on good collaboration with him.[616] Finally, Settele was able to note somewhat triumphantly in his diary on November 24 that de Crollis's excerpt had now appeared with Anfossi's imprimatur after all.[617]

As this farce unfolded, naturally, Olivieri too played a part. After having refuted Anfossi's arguments repeatedly in a lengthy, detailed opinion in August 1822, he himself now went on the offensive and recommended that the congregation should now finally remove the names of Copernicus, Zúñiga, and Foscarini from the Index.[618] The consultors dealt with this suggestion in their *Feria Secunda* meeting on September 2, 1822, resulting in a unanimous vote to recommend

613 See minutes of the meeting on September 11, 1822, in Brandmüller and Greipl, *Copernico, Galilei e la Chiesa*, 427–473.

614 See Anfossi to Odescalchi, September 25, 1822, in ibid., 429. The reply, however, consisted only of the passages from the relevant 1616 and 1620 decrees that seemed to Anfossi to be conclusive, along with the oft-quoted statement by Boscovich, "*nobis hic Romae*" (for us here in Rome). It should be noted that Olivieri had to come to a negative decision about this letter, too.

615 See Odescalchi to Turiozzi, September 26, 1822, in Brandmüller and Greipl, *Copernico, Galilei e la Chiesa*, 430–431.

616 See Settele, October 2, 1822, in Maffei, *Giuseppe Settele*, 412.

617 See Settele, November 24, 1822, in ibid., 414.

618 See Brandmüller and Greipl, *Copernico, Galilei e la Chiesa*, 412–426.

to the Congregation of the Index that it proceed accordingly with a new edition of the Index.[619] When this vote was presented to the congregation of cardinals at their *Feria Quarta* meeting on September 11, they decided that the request should be granted.[620] However, misgivings seem to have cropped up immediately, because the following *Feria Quarta* meeting on September 18 ordered that the execution of the decree dated September 11 be suspended. Meanwhile, they should investigate whether the works mentioned had been found objectionable exclusively on account of their Copernican teaching and not for other reasons also.[621]

Thus Olivieri was assigned to examine whether the works of Copernicus, Foscarini, and Zúñiga had been condemned in their day merely on account of their heliocentrism or for other reasons, too.[622] The fact that they had found in the archives first the relevant act of the Congregation of the Index in the original manuscript made it easy for him to answer this question, in the negative. As far as Copernicus was concerned, Olivieri could cite Ingoli, since the latter had compiled in 1620 the list of corrections required if the prohibition was to be lifted.[623]

Regarding the other works, he himself set to work once again. He found both Zúñiga's commentary on Job[624] and also Foscarini's booklet[625] in the original editions in the Biblioteca Casantense, which was familiar to him as a Dominican; they were by no means stored in the cabinet reserved for forbidden books. They merely

619 See minutes of the meeting on September 2, 1822, in ibid., 426.

620 See minutes of the meeting on September 11, 1822, in ibid., 427–427.

621 See minutes of the meeting on September 18, 1822, in ibid., 428.

622 See Olivieri's expert opinion, early 1823, in ibid., 440–480.

623 See Favaro, *Le Opere di G.* Galilei, 5:403–412; cf. GT74-75 = around footnotes 91-92 of Part I, chapter 2.

624 Call number: Biblioteca Casanatense KK VII 32.

625 Call number: Biblioteca Casanatense Vol. Misc. 75/2.

displayed handwritten corrections, as was customary for editions in those days — even the work by Foscarini, which had not been forbidden merely "*donec corrigatur*" (until it was corrected). This is characteristic of the way in which the 1616 and 1620 decrees were applied. Olivieri does not forget to mention that the first edition of Diego Zúñiga's commentary on Job, which appeared in 1584 in Toledo, was dedicated to King Philip II of Spain, and an edition produced in 1591 in Rome was dedicated to Pope Gregory XIV, which would not have been possible without their permission.

In the first work he found only one passage that came into question for an investigation. It was the commentary on chapter 9, verse 6: "He shaketh the earth out of her place, and the pillars thereof tremble." Zúñiga in fact interpreted this verse in the Copernican sense, although, as Olivieri accurately notes, it expresses only the divine omnipotence over all of nature. Moreover, in the copy from the Casanatense library it was the only sentence crossed out, which showed him that the book had been forbidden solely because of it.[626]

On the other hand, Olivieri had to grapple more extensively with Foscarini, whose booklet, as mentioned, had likewise been found in the Casanatense. The copy in question was marked up with numerous marginal notes in the handwriting of its first owner.[627] In it Olivieri found one pretext or another for possible criticism, and therefore he subjected it to a more careful examination, whereby he reports some of the aforementioned marginal notes verbatim, so as to refute them at once.[628]

[626] The sentence in question is on page 205 of the book.

[627] The previous owner of the work was Patritius de Olivis (?), Medicus et Physicus, as noted on the frontispiece of the copy in the Casanatense.

[628] See, for example, Brandmüller and Greipl, *Copernico, Galilei e la Chiesa,* 442–443.

With that, he thought, his work might really be done. However, if these three aforementioned works were stricken from the Index, then it should be done also for Kepler's *Epitome* and Galileo's *Dialogo*. They too had been forbidden only on account of their heliocentrism. Now the prohibition must finally be abolished — not only because it had been imposed only on account of Kepler's heliocentrism, but also considering the admirable corrections of his theory in his later works and on account of the truly Catholic attitude with which he accepted the prohibition in his day.[629]

Now Olivieri still had to deal with Galileo. His *Dialogo*, too, had been forbidden exclusively because of its Copernican conviction. The blameworthy feature was that in connection with the grant of permission to print the *Dialogo* he had concealed the fact that he had been forbidden by order of the pope to continue speaking on these questions. He had transgressed this prohibition by publicly defending Copernicus in his *Dialogo*. The permission for the Padua edition, however, already signified a "*perdono*" (pardon) for this transgression.[630]

In Olivieri's opinion, the omission of the general ban of heliocentric books by the 1758 Index spoke in favor of lifting the prohibition on the *Dialogo*. This decision could be applied without further ado to the books of this sort listed by name on the Index, and therefore also to the *Dialogo*. A further argument in favor of this was provided by the fact that the various other writings of Galileo, which expressed the same ideas, were never forbidden by name, but rather circulated freely, whereby Olivieri was referring both to the writings about sunspots and also to the letters to Castelli and the Grand Duchess Mother Christina.

[629] See Leopold Schuster, *Johann Kepler und die grossen kirchlichen Streitfragen seiner Zeit* (Graz: Moser, 1888), 131–135. Naturally it is out of the question that Kepler was Catholic (ibid., 202–232).

[630] See Brandmüller and Greipl, *Copernico, Galilei e la Chiesa*, 448–449.

But now Olivier had to overcome an obstacle that he himself had set up. Recall that he had repeatedly argued that the former prohibitions of the books now in question had been imposed rightly on account of their inherent absurdities, from which the heliocentric theory had now been liberated. However, he dealt easily and convincingly with this difficulty by pointing out that by his day these works were of historical interest only.

After presenting once again the course of events concerning Settele, and also pointing out that an excerpt from his work had appeared in the *Giornale Arcadico* — indeed, with Anfossi's permission[631] — he posed the additional question: How else can the contradiction be remedied between a continuance of the prohibition of Copernicus and his allies by name and the imprimatur granted to Settele?

Anfossi had been the one to raise this question again, although Olivieri had shown that the contradiction between the earth's movement taught by Galileo and the real meaning of the oft-cited Scripture passages about the earth's immobility really had existed in 1616 and 1633, but now was removed. With his "contradiction," therefore, Anfossi was contradicting the evidence. Nevertheless, Olivieri says, it is still difficult to explain — and there is no denying it — that the execution of the 1758 decree of the Congregation of the Index was incomplete.

On the other hand, though, we know that in the new edition of the Index in 1758 the intention was to reduce the volume of the Index by summarizing many individual prohibitions in general prohibitions (thus "*libri omnes, qui...*" [all books that...]). However, if these five books remained on the Index, although the general prohibition under which they fell had been abolished, then

[631] See ibid., 456–457.

we can assume that they too were no longer forbidden. Besides, Anfossi with absolute contempt dismisses all progress in astronomy, just as he ignores the arguments advanced by Olivieri. Instead, he stereotypically repeats his accusation that heliocentrism contradicts Sacred Scripture, as the Church Fathers explained it. This statement gives Olivieri the opportunity to emphasize again that the theses formulated by the qualificators of the Holy Office in 1616 by no means had led to papal definitions, as Anfossi groundlessly maintained now as before. Plainly, he lacks the courage to admit this. Moreover, this is true also of the Galileo judgment in 1633 — an argument that he develops "*per longum et latum*" (in its full length and breadth). His conclusion: not a single one of Anfossi's objections can be upheld.

Olivieri now concludes his expert opinion with a recommendation about how to proceed in practice, if they intend to strike these five titles from the Index where they are mentioned by name. It can be accomplished most easily by replacing the five pages on which the titles under discussion appear in the unbound remnant of the 1819 edition of the Index, with a new printing that no longer contains them. The title page could also be replaced by one bearing the correct year of publication. A new foreword could be added. In this way the recommended step would not have to wait for a new edition of the Index.

But now the illness and death of Pius VII — he died on August 20, 1823 — caused the business to retreat into the background. Not until November did the consultors of the Holy Office discuss the matter again in the *Feria Secunda* meeting on November 10, 1823; they decided that Olivieri should deal in a further opinion with the objections that had been presented in the session or with other conceivable counterarguments. The text of this opinion also allows us to reconstruct the course of the discussion on November 10, for which no other sources exist.[632]

According to the document, the difficulties raised at the meeting no longer referred to Copernicus, but rather to Zúñiga. Again it was a hermeneutical question, since Zúñiga had maintained that the book of Job explicitly spoke about the movement of the earth. Now Olivieri correctly shows, among other things, that these remarks by no means referred to astronomical movement. Rather, they were meant to express in poetic language God's omnipotence over all creation. His conclusion: the objectionable passage in the Job commentary is no obstacle at all to lifting the prohibition.

The following discussion about Foscarini proved to be more difficult. At stake here were systematic aspects of the Church's teaching authority. Indeed, Foscarini had written that the pope is inerrant only in questions of faith and the salvation of souls, but may err in practical judgments, philosophical speculations, and other matters that do not concern the salvation of souls. Now, considering that this statement had caused offense, Olivieri proceeds very carefully in this matter; nevertheless he emphasizes that this simply repeats general Church teaching, whereby he cites Melchior Cano. Besides, he points out that Foscarini wrote *before* the 1616 censures and furthermore explicitly and solemnly declared his fidelity to the Magisterium. However, he continues, can we indeed say that the Church can err in philosophical speculations? *Tutto sommato* (All told), there is nothing objectionable about that, when we apply the thesis to the examples listed — the antipodes or the inhabitants of the *Zona Torrida*. After all, we are talking here about objects of natural rational knowledge.

A second objection had been raised by the consultors to the exegesis by Foscarini, who saw in the seven-branched candelabra of the

[632] See Olivieri's opinion dated November 1823, in ibid., 462–480.

Temple a reference to the seven planets. Olivieri replied to this with a defense of allegorical interpretation, pointing out not only the pulpit style in the early seventeenth century but also several examples from the works of the Church Fathers and also of the famous exegete Sixtus of Siena,[633] which had proceeded in a similar fashion. So this objection too was invalidated.

Now, though, Olivieri dealt with Kepler more thoroughly than with all the other authors, because he was plainly quite fascinated by his *Epitome*. Eighteen titles by the astronomer were in the Casanatense library. What particularly impressed him was Kepler's pious, respectful way of dealing with his topic. He also admired his excellent knowledge of Greek and especially the fact that Kepler even understood Hebrew. Two of Kepler's ideas had now met with misgivings in the Holy Office. First, that Kepler saw in material creation a likeness of the Divine Trinity, and second, that he viewed the stars as having souls.

On the first issue, Olivieri says about all the passages that are relevant in this context, which he himself cites at length and verbatim, that they contained the expression of an imagination rather than rational arguments. Yet they contained nothing inconsiderate of the faith or inconsistent with sound theology. Moreover, Kepler's manner of speaking removes all doubt that he is talking about similarities, images, and analogies. Augustine and Thomas Aquinas had done this, too, when they spoke about the world's relation to God.

[633] Sixtus of Siena (1520–1569), born a Jew, converted in his youth and became a Franciscan, then a Dominican. Michele Ghislieri (later Pius V) rescued him from condemnation by the Inquisition on account of heretical teachings. From then on he was an outstanding, unquestionably orthodox exegete and an important preacher. He left behind an eight-volume work, *Bibliotheca sancta* (Venice, 1566), which went through many editions. See A. Schmitt, in *KL2* 9 (1899): 384–386; J. Molitor, in *LThK2* 9 (1964): 812.

As far as the animation of the stars is concerned, Olivieri forthrightly admits that such an idea sounds odd to contemporary ears; it was otherwise in the past, however, since the time of Origen. Ambrose and Augustine spoke about it, and Thomas opined that whether or not the stars had souls was not an article of faith. In any case, though, the philosophers of antiquity — Plato, Thales, Pythagoras, the Stoics, and Aristotle — had assumed the animation of the stars, as did Philo and Clement of Alexandria even before Origen.[634]

Huetius had already demonstrated this and so had taken up Kepler's defense.[635] Nonetheless Olivieri himself now cites a series of relevant passages by Kepler, to show that in Kepler's writings all these ideas ended in a hymn to the Creator. There is really no obstacle whatsoever to striking Kepler's work from the Index, he concludes.

Although the discussion about Kepler had proved to be rather long, Olivieri goes on to say that he can be much briefer regarding Galileo's *Dialogo*. First, there was the objection that, with the character of Simplicio, Galileo had held Urban VIII up to ridicule. Olivieri comments that Galileo's enemies circulated that rumor in order to harm him.[636] The real objections to the *Dialogo* can be derived from Ingoli's remarks.[637] Ultimately, these did not originate with then-Cardinal Barberini, but rather were common property of the Peripatetic school. Moreover, Galileo's conduct with regard to the Holy Office in connection with the printing of the *Dialogo*, or with the necessary imprimatur,

634 The notion that the stars are moved or inhabited by angels or astral spirits goes back to Plato and Aristotle and was adopted by Scholasticism. Paracelsus had similar views also. See H. M. Nobis, in *LMA* 1 (1980): 1132–1133.

635 Also known as Pierre Daniel Huet (1630–1721), a convert, bishop of Avranches, a famous polymath, historian, author of numerous scholarly works. See F. X. v. Funk, in *KL2* 6 (1889): 335–338.

636 Here Olivieri is in fact striving to whitewash Galileo.

637 See Favaro, *Le Opere di G. Galilei*, 5:403–412.

had been an object of the consultors' criticism. Olivieri maintains, "*Ammetto pertanto le accuse, e le credo vere; ma la conseguenza mi sembra in allegerimento della colpa del Dialogo, poichè sebbene non si parlasse in pubblico se non di questa; in realtà però se ebbero in mira altre delinguenze maggiori.*" (I therefore admit the charges, and I think that they are true; but it seems to me that the result is a mitigation of the fault of the *Dialogo,* since although in public people spoke about nothing else but this, in reality they had other greater crimes in view.)[638] The third point of criticism was that, according to the judgment of serious historians, Galileo thought that the Copernican theory could be proved by Sacred Scripture and wanted it to be raised to the status of a dogma and its antithesis condemned as a heresy. Olivieri replies that the only reference that he could find to a claim that Sacred Scripture favors the Copernican teaching was in Galileo's letter to Castelli, which however was not condemned! Thus there is not one single reason to leave the *Dialogo* on the Index.

But now other misgivings concerning matters of principle had also been expressed about the intention to strike the aforementioned works from the Index. Some were worried about how the Curia would save face in this case. Olivieri's answer was "*che la fa anzi buonissima*" (on the contrary, this puts the best possible face on it).[639]

A second misgiving: Wouldn't it disavow the esteemed consultors who in 1758 left these books on the Index if we now deleted them? Olivieri opines that back then, since they had thousands of books to deal with, they could not worry about individual cases. If in 1758 they had dealt with Galileo as intensively as we have done today, no doubt they would have come to the same conclusion.

638 Olivieri's expert opinion dated November 1823, in Brandmüller and Greipl, *Copernico, Galilei e la Chiesa,* 478.

639 Ibid., 479.

And finally there are other books which were condemned for the same reason as these, which should therefore also now be stricken from the Index. Olivieri says that he believes that someone who says that is correct. He himself, however, found none. And the most famous work that advocates the Copernican teaching is, after all, Newton's *Principia*. But this was never condemned!

With that Olivieri concluded his argumentation with a final persuasive reference. But even this expert opinion was unable to cause a breakthrough. When the *consulta* of the Holy Office debated the matter again on December 1, 1823, again the only result was a "*dilata*" (postponement): Fr. Cappelari — the future Gregory XVI — and Fr. Garofalo were to make their remarks about it, and the consultors should discuss it once more.[640]

There is no documentary evidence for the further deliberations; the sources now dry up. It is certain, however, that the new Index issued in 1835 by Gregory XVI no longer contained the aforementioned works.[641]

After that we have only sporadic reports. For example, Settele's student Purgotti wrote a treatise against Anfossi,[642] and both Olivieri and Settele himself suggested improvements to it.[643] At around the same time that Purgotti was at work, Olivieri sent a circular letter to

[640] See minutes of the meeting on December 1, 1823, in Brandmüller and Greipl, *Copernico, Galilei e la Chiesa*, 481. The texts of their expert opinions unfortunately could not be found in the archive.

[641] See Reusch, *Der Index der verbotenen Bücher*, II/2, 878–883.

[642] The seventy-four-page work by Sebastiano Purgotti (1799–1879), who later became an important mathematician, is entitled *Riflessioni sopra un opuscolo che porta per titolo "Se possa difendersi ed insegnare non come semplice ipotesi ma come verissimo e come tesi la mobilità della terra, e la stabilità del Sole da chi ha fatto la professione di Fede di Pio IV"* (Pergola, 1823). The imprimatur for it was granted on July 14 or August 5, 1823.

[643] See Settele, June 5, 1823, in Maffei, *Giuseppe Settele*, 414–415. On October 2, 1823, Settele reports that the work had appeared in print (ibid., 416).

the inquisitors of several Italian cities, in which he informed them that Settele's *Astronomy* had been published with the permission of the Holy Office.[644] All this took place in the year 1823. Six years later, when a Spanish bishop inquired with the Holy Office whether a Catholic was allowed to advocate the Copernican system, the dicastery replied to this altogether anachronistic inquiry by forwarding without comment the relevant decrees in the Settele affair.[645]

In Rome Settele's conflict with Anfossi was still recalled often in jest. For example, the former expert Cappellari, now Pope Gregory XVI, during an audience in 1833, grinned at Professor Settele and asked, "*Gira la testa o gira la terra?*" and added, "*Ma lasciate che giri la terra, basta, che non girino le teste!*" (Does your head turn or does the earth turn? Just let the earth spin; it is enough if heads do not spin, too!)[646]

[644] See Settele, June 18, 1823, in ibid., 415. Settele refers here to a communication from Olivieri. No trace of it was found in the archive of the Holy Office.

[645] See Settele, March 29, 1829, in ibid., 421. Here too Settele reports a communication from Olivieri. Perhaps the entry in the Collectanea of the Aso 1800–1849, no. 59, "Hispalen, Censura Librorum 1826–1830 no. 4" refers to this proceeding. The archival material itself proved to be untraceable.

[646] See Settele, August 28, 1829, in ibid., 421.

EPILOGUE

THE USUAL JUDGMENT about the Galileo case is, as we described earlier, unequivocally to the detriment of the Church. It holds that the Church condemned a finding of the natural sciences which immediately was to become generally accepted and today is the common property of all, and thus not only put itself in the wrong for all time but also suffered a general dwindling of trust, which became manifest when many scientists left the Church. Rigid dogmatism, fixated on keeping power, won out over the free human intellect; a victory, but what a Pyrrhic victory!

Nevertheless, things are not that simple. At the moment when the decision was about to be made in Rome, only a prophet could have known that Galileo would initially — but only initially — be corroborated by further developments, while the Curia would be declared incompetent. Nor were people aware of the paradox that each of the opponents was in the wrong on his own terrain. Although Galileo's propaganda for Copernicus was justified later on by Newton, Bradley, and others, he himself erred with regard to the conclusiveness of the arguments that he had advanced in favor of the Copernican system. Today there is no longer the slightest doubt about this, and his critical contemporaries already knew this, too. There are good reasons to assume that, for example, the Jesuit astronomers Grienberger and Scheiner thought so, and somewhat later without any doubt Riccioli, too, as well as other Copernicans. The fact that they did not openly profess Copernicanism was subsequently interpreted as comfortable conformism or a blind, pious obedience, if not as hypocrisy. Both explanations miss the heart of the matter: They may have been intuitively convinced by the heliocentric system, as Galileo was, but saw no strict proof for it. The fact that Galileo's arguments did not prove what they were supposed to prove did not remain concealed from them. The Jesuits, however, were the advisors of the Curia on matters of natural science. Moreover, as was shown, among astronomers there were staunch opponents of Galileo well into the eighteenth century. Certainly, in one case or another, intellectual immobility or narrow-mindedness may have been the cause of this. Overall, however, the prevailing factor may well have been the skepticism that is inherently part of the theory of science, which in our day—and evident already in Thomas Aquinas—is again conscious of the conditions for and limitations of human knowledge; being averse to all naive epistemological optimism, it emphasizes the provisional character of scientific knowledge.

The result of this is the grotesque fact that, in the final analysis, the Church, which is often called to account for erring in this question, was correct in the field of natural science (which for it is more remote) when it demanded that Galileo advocate the Copernican system only as a hypothesis. As early as 1908 the physicist Pierre Duhem therefore made from his perspective on the Galileo trial the astonishing observation that "logic was on the side of Osiander, Bellarmine, and Urban VIII, and not on Kepler's and Galileo's; the former recognized the real significance of the experimental method, while the latter misunderstood it."[647] And now the reasons for this: "Assuming that the hypotheses of Copernicus could explain all known [celestial] phenomena, from that one could conclude that

[647] Cited by Crombie, *History of Science*, 451.

they were possibly true, but not that they were necessarily correct. For in order to legitimize the latter conclusion, one would have to prove that no other system is conceivable that explains the phenomena just as well. This latter proof, however, was never offered."[648] But that was precisely the state of the cosmological question when the Galileo case was being discussed in Rome. Also when considered from the standpoint of today's theory of science, the Church's reservations about heliocentrism appear well founded. Thomas Aquinas was the one who had formulated the principle that the Roman theologians had followed, and in that way they avoided falling into a naive scientific optimism.

Now of course the Copernican system was condemned in 1616, and Galileo's *Dialogo* in 1633, not because the Church considered the heliocentric system false and the cosmology of Ptolemy or of Tycho Brahe true. The Roman "no" to Copernicus and Galileo was based primarily instead on the assumption that the heliocentric system contradicts Sacred Scripture. Upon closer inspection, this assumption becomes quite plausible, given the presuppositions at that time.

The various Bible passages that scholars used to quote against the teaching about the earth's movement speak in fact about the earth as the unshakably firm arena of human life. For all the coming and going that occurred on it every day, the earth remained where and what it was. Finally, without knowledge about the weight of the air and about gravitation, which was not yet available at that time, no one could imagine, assuming that the earth moved both around its

[648] Cited ibid. Interestingly enough, that is precisely the objection that Cardinal Barberini (Urban VIII) raised, which Galileo then placed on the lips of his "Simplicio" and made fun of; it goes back to Thomas Aquinas. See above, GT115 [not "112"] = beginning of Part 1, Chapter 3, after footnote 5.

own axis and also around the sun, how any human being could remain on his feet, how a house or a tree could be left standing, if the earth was hurtling through the universe at a furious pace.

The claim that the earth moves around its own axis, given the presuppositions then, therefore really did contradict the literal meaning of Sacred Scripture, which speaks about the unshakable steadfastness of the earth. The fact that the Roman theologians in the years of the Galileo case could not imagine how extensively these presuppositions would change within a half a century cannot be held against them as an error.

Instead, we can criticize the fact that they tried to answer a physical-astronomical question with theological methods in the first place. We would have to describe such criticism, however, as anachronistic, because it utterly disregards the aforementioned holistic vision of the Baroque period; against this background the argumentation of the Roman theologians appears to be quite understandable.

What can be criticized is the fact that Galileo's theological opponents were not capable in this case of interpreting the literal sense of Sacred Scripture adequately, by adopting the principles of interpretation that had already been applied by Cajetan, or—like other contemporaries—by taking into account the possibility that Sacred Scripture was employing colloquial speech at the places indicated.[649] This, however, can of course be regarded as an error of the Inquisition.

And Olivieri's solution after two hundred years? At first it might seem that he had relinquished the position of Aquinas under the

[649] See, for example, Loretz, *Galilei und der Irrtum der Inquisition,* 101–113; Carlo Maria Martini, "Gli esegeti del tempo di Galileo," in *Quarto centenario della Nascità di Galileo Galilei,* Pubblicazioni dell'Università Cattolica del Sacro Cuore, Contributi III/8 (Milan, 1966), 115–124; Giovanni Leonardi, "Verità e libertà di ricerca nell'ermeneutica biblica cattolica dell'epoca Galileiana e attuale," *Studi Patavina* 29 (1982): 597–635.

impression of scientific progress. But Olivieri too, and then the Holy Office along with him, did not maintain that the earth's movement and heliocentrism were indisputable truths — although by then the professional world was convinced of it, and public opinion no less so. Olivieri's reasoning merely shows that someone can teach this astronomical interpretation without contradicting the Catholic faith. This restraint immediately proved to be all too justified, for further research has long since made the system of Copernicus, Galileo, and Newton obsolete. And this same development confirms once again the Thomistic methodological skepticism of the Roman theologians in 1616. In making this statement, therefore, the Holy Office had kept strictly within the boundaries of its theological-scientific and also ecclesiastical-magisterial competencies.

The result is the paradox that Galileo erred in natural science, and the Curia in theology, while the Curia was right in natural science and Galileo in biblical exegesis.[650]

The question now is: How should this state of affairs be judged? Isn't this an impressive demonstration of the historicity of man and of human knowledge? Whenever we claim today to have progressed in our knowledge, compared to past generations, let us nevertheless acknowledge also the possibility, indeed the certainty, that the measure of knowledge that we have attained will also be surpassed. No one can say how many more "Copernican revolutions" mankind will experience. But just as we today are obliged to resolve our present-day problems and questions on the basis of what it is granted to us to know, so too were Galileo's followers and opponents. We must therefore allow them the same thing that we too are compelled to claim for ourselves: the right to error, or more precisely, to the provisional character of our knowledge and thus the right to make progress in it.

[650] See Langford, *Galileo, Science and the* Church, 78.

How unbearably, inhumanely pharisaical it would be to accuse people in earlier times for not already knowing everything that is self-evident to us today!

In the field of natural scientific research and discovery, this is now admitted, too. Of course, when it is a matter of acknowledging religious truth and the proclamation of it by the Church, most people judge otherwise! The Church itself, however, never made the claim to be in fully conscious possession of the whole truth from the very beginning. It has always been aware that it must first be led into all truth by the Holy Spirit dwelling within it — as the Gospel of John says.[651] With this "leading into," though, the knowledge of divine revelation is already defined as a process, and the Gospel significantly says nothing about its course or duration. For here the concept of dogmatic development too has its theological *locus*. Vatican Council II formulates it accordingly: "This tradition which comes from the Apostles develops in the Church with the help of the Holy Spirit. For there is a growth in the understanding of the realities and the words which have been handed down.... For as the centuries succeed one another, the Church constantly moves forward toward the fullness of divine truth."[652]

Now, of course, someone will immediately raise the objection that the Church does claim infallibility in matters of Faith. Certainly it does, and it correctly bases this claim on Sacred Scripture. This infallibility, however, means only that the Church is preserved from error in proclaiming dogmas that are binding for all the faithful. On

[651] "But the Counselor, the Holy Spirit, whom the Father will send in my name, he will teach you all things, and bring to your remembrance all that I have said to you" (Jn 14:26:). "I have yet many things to say to you, but you cannot bear them now. When the Spirit of truth comes, he will guide you into all the truth" (Jn 16:12–13: ; RSVCE).

[652] Vatican Council II, Dogmatic Constitution on Divine Revelation *Dei verbum* (November 18, 1965), no. 8.

the other hand, infallibility does not mean full knowledge of the truth of revelation. That is possible only in the vision of God in the next world. Yet, when the Church claims infallibility for its doctrines, it means that, as stated, they contain no error, although they can be and need to be formulated more completely and deeply.

Obviously, as already mentioned, this claim is not made for every single statement of any ecclesiastical authority. Only the pope and an ecumenical council are qualified to exercise infallible teaching authority, which is therefore definitively binding. Furthermore, in order to qualify as such, their statement needs to be in the appropriate form. Consequently, in many of those cases in which ecclesiastical spokesmen do not speak with that binding force, error is not absolutely ruled out as a matter of principle. And the Galileo case is such a case; no dogma was ever formulated in connection with it. It may be regrettable that neither the cardinals who were responsible nor Paul V and Urban VIII recognized the nullity of the contradiction that was asserted between Copernicus and the Bible, as we can see today. But we cannot dispute their right to this error. Since in the apparent dilemma they considered the sacrosanct character of the Bible to be a higher good and therefore more in need of protection than an astronomical cosmology which was not even strictly proved, they did their duty as protectors of the revealed Faith. This should not be held against them, but at the same time it was the tribute that they paid to the historicity of human knowledge. Only someone who was sure of being personally free of the duty to pay this tribute could set himself over human knowledge as its judge.

It is all the more astounding that the few official statements of the Congregation of the Index dated 1616 and 1620 and of the Holy Office dated 1633 were formulated in such a way that two hundred years later they allowed each and every ecclesiastical obstacle to the acceptance of the earth's movement to be removed, without disavowing the decisions from the time of Galileo.

For the Church, the "Galileo case" has been closed since 1820.

The dialogue between the natural sciences and the Church, on the other hand, seems new and fraught with significance for the future.

For more sensitive thinkers such as Ernst Jünger, who considered himself a natural scientist, and not just because of his education, this dialogue had already been prompted by the experience of World War II, a catastrophe that he viewed as the extreme consequence of modern thinking. In the final days of the war he therefore made the demand: "We must find our way back along the path that Comte marked out: from science via metaphysics to religion. Of course, it was less laborious downhill."[653] Of course this does not mean that Jünger wanted to replace natural science with metaphysics and religion. Rather, he is speaking about "modern physics, which as a *scienza nuova* [new science]" — and this is a clear allusion to Galileo's *Discorsi* — "sets free all the forces of chance and will experience adequate restriction only through the queen of sciences — theology."[654]

It took more than three decades for these insights to become widely known and accepted. Comte's positivism commands important bastions even today. As late as 1968 a leading physicist could say, "It is a shame, but the Catholic Church and the modern natural sciences are worlds apart."[655]

A declaration directly contradicting this statement was issued by twelve Nobel Prize winners on December 22, 1980, in Rome in an

[653] "Die Hütte im Weinberg," May 1, 1945, in Ernst Jünger, *Strahlungen II*, Sämtliche Werke, Tagebücher III (Stuttgart, 1979), 422.

[654] *Zweites Pariser Tagebuch*, March 7, 1944 (ibid., 234).

[655] Brüche, "Das Angebot des Kardinals," 363.

address to Pope John Paul II. Like Ernst Jünger at the end of World War II, they emphasize the necessity of ethical restrictions for natural scientists and technicians — for the sake of mankind's future. Therefore they demand that this outmoded wall which until now has separated natural science and religion be torn down, and they continue by saying, "In particular we recognize that the Catholic Church is in a unique position to provide moral guidance on a world-wide scale."[656] They therefore greatly appreciate the willingness of the pope to discuss with them the problems of humanity in the light of modern science.

This is a document of epochal importance, which in principle builds a bridge over the oft-lamented gulf between natural science and religion and, knowing their joint responsibility for the future of mankind, opens wide the doors for a productive dialogue.

But now, in all probability, the declaration of the twelve Nobel Prize winners will have a fate similar to that of the Copernican system: It will have to raise public consciousness in a long process that makes the contents of this declaration the common property of those who pursue the natural sciences, and this process will be determined by presuppositions and factors similar to those that affected the reception of heliocentric cosmology. Knowledge about the necessity of religiously and theologically substantiated norms will intervene in the thinking and living of an individual scientist no less than the "Copernican revolution" did in its time. Patience will be necessary, therefore, until the spark has caught fire everywhere.

Galileo himself could volunteer to lead his fellow scientists of today in such a revolution; after all, both in his correspondence and especially in his letter to the Grand Duchess Christina, he showed and

[656] "Declaration by Twelve Nobel Prize Winners," delivered on December 22, 1980, in Rome. (See page 337.)

also argued intellectually that Catholic belief and research in the natural sciences can be brought into fruitful harmony in one and the same person.[657] Whereas until now Galileo was a figure symbolizing the conflict between science and faith, may his name in the future testify to the possibility and productivity of harmony between the two.

[657] The view expressed by Giorgio Spini, "The Rationale of Galileo's Religiousness," does not bear closer examination of the source material. A much more accurate account is Pietro G. Nonis, "Galileo e la Religione," in *Quarto centenario*, 125–152. The best presentation of Galileo's Catholic belief is offered by Olaf Pedersen, "Galileo's Religion," in *The Galileo Affair*, 75–102.

Declaration by Twelve Nobel Prize Winners

delivered on December 22, 1980, in Rome

J. Dausset Medicine France	J. Eccles Medicine Australia	F. A. von Hayek Economics Great Britain
H. A. Krebs Medicine Great Britain	I. Prigogine Chemistry Belgium	M.H.F. Wilkins Medicine Great Britain
C. de Duve Medicine Belgium	E. O. Fischer Chemistry Australia	L. R. Klein Economics U.S.A.
S. Ochoa Medicine U.S.A.	C. H. Townes Physics U.S.A.	R. S. Yalow Medicine U.S.A.

Rome, December 22, 1980

We Nobel Prize winners share with Alfred Nobel his great concern that science should benefit humanity.

Science has brought great benefits and we look forward to it continuing to do so.

Nevertheless, scientific knowledge has sometimes been applied in very undesirable ways, for example in war, and beneficial applications may have unexpected side-effects which are undesirable.

Furthermore, the intellectual enlightenment that science has provided has changed humanity's view of itself and its place in the universe. This has tended to leave human beings spiritually deprived and in a moral vacuum.

We believe that scientists must possess ethical sensitivity, and we are eager to break down the traditional separation — or even opposition — of science and religion.

The churches clearly can have a special role in trying to achieve this aim; and in particular we recognize that the Catholic Church is in a unique position to provide moral guidance on a world-wide scale. We therefore greatly welcome the opportunity provided by Nova Spes[658] for us to come together to discuss the position of science in our culture, and we greatly appreciate the willingness of Your Holiness to discuss with us the problems of humanity in the light of modern science.

[658] Nova Spes is an international foundation founded by Cardinal Franz Koenig.

Bibliography

Only titles that have been mentioned at least twice in the notes are listed here.

Accademia dei Lincei. *Galileo Galilei, Celebrazioni del IV. Centenario della Nascita.* Rome, 1965.

Accademia Patavina di scienze, lettere ed arti. *Scritti e discorsi nel IV. Centenario della nascita di Galileo Galilei.* Padua, 1966.

Atti del Convegno di Studio su Pio Paschini nel Centenario della Nascita 1878–1978, ed. Deputazione di Storia patria per il Friuli. Udine, 1979.

D'Addio, Mario. *Considerazioni sui processi a Galileo*, Quaderni della Rivista di storia della Chiesa in Italia 8. Rome, 1985, 237.

———. *Il caso Galilei, Processo — Scienza — Verità.* Rome, 1993.

Aubert, Roger. "L'État actuel de l'affaire Galilée," in: *Colloque d'Histoire des Sciences*, Université de Louvain, Recueil de Travaux d'Histoire et de Philologie VI/9. Louvain, 1976, 151–163.

Bangen, Johann Heinrich. *Die Römische Curie, ihre gegenwärtige Zusammensetzung und ihr Geschäftsgang: Nach mehrjähriger eigener Anschauung dargestellt.* Münster, 1854.

Bidolli, A. P. "Contributo alla storia dell'Università degle Studi di Roma, La Sapienza durante la Restaurazione," in: *Annali della Scuola Speciale per gli Archivisti e Bibliotecari dell'Università di Roma* 19/20 (1979/1980): 71–110.

Blumenberg, Hans. *Die kopernikanische Wende*. Frankfurt, 1965.

Boffito, Giuseppe. *Scrittori Barnabiti o della Congregazione dei Chierici Regolari di San Paolo (1533–1933), Biografia, Bibliografia, Iconografia* II. Florence, 1935.

Borselli, L, Poli, Ch., and Rossi, P. "Una libera Comunità di Dilettanti a Parigi del '600," in: *Cultura Popolare e Cultura Dotta nel Seicento*, Atti di Convegno di Studio di Genova 23–25. Nov. 1982. Milan, 1983.

Boscovich, Ruggero Giuseppe. *De Aestu Maris Dissertatio*. Rome, n.d. [1747].

Brandmüller, Walter. *Galilei e la Chiesa, ossia il diritto ad errare*, Scienze e Fede 4. Vatican City, 1992.

Brandmüller, Walter and Greipl, E.J. *Copernico, Galilei e la Chiesa, Fine della controversia (1820), Gli atti del Sant'Uffizio*. Florence, 1992.

Brodrick, James. *Robert Bellarmine: Saint and Scholar*. London, 1961.

Brüche, Ernst. "Das Angebot des Kardinals: Aus der Eröffnungssitzung der 18. Lindauer Nobelpreisträgertagung," in: *Physikalische Blätter* 24 (1968): 358–363.

Cinti, Dino. *Biblioteca Galileiana raccolta dal principe Giampaolo Rocco di Torrepadula*, Biblioteca bibliografica italica, Contributi 15. Florence, 1957.

Colloque d'Histoire des Sciences, Université de Louvain, Recueil de Travaux d'Histoire et de Philologie IV/9. Louvain, 1976.

Crombie, Alastair Cameron. *Von Augustinus bis Galilei, Die Emanzipation der Naturwissenschaft*. Cologne-Berlin, 1965.

Del Re, Niccolò. *La Curia Romana, Lineamenti storico-giuridici*, Sussidi Eruditi 23. Roma, 1970.

I Documenti del processo di Galileo Galilei, ed. S. M. Pagano and A. G. Luciani, Pontificiae Academiae Scientiarum Scripta Varia 53. Vatican City, 1984.

Drake, Stillman. *Galileo at Work: His Scientific Biography*. Chicago-London, 1978.

——. "Galileo and the Church," in: *Rivista di Studi Italiani* 1 (1983): 82–97.

Encyclopédie ou Dictionnaire raisonné des Sciences, des Arts et des Métiers, IV. Paris, 1754.

Eszer, Ambrosius K. "Niccolò Riccardi, O.P., il 'Padre Mostro,' " (1585–1639), in: *Angelicum* 60 (1983): 428–461.

Ferraris, Lucius. *[Prompta] Bibliotheca canonica, juridica, moralis, theologica ... in octo tomos distributa*. Rome, 1759–1761, or Bonn, 1763.

Ferretto, Giuseppe. *Note storico-bibliografiche di Archeologia cristiana*. Vatican City, 1942.

Feyerabend, Paul K. *Der wissenschaftstheoretische Realismus und die Autorität der Wissenschaften, Ausgewählte Schriften* I, Wissenschaftstheorie, Wissenschaft und Philosophie 13. Braunschweig-Wiesbaden, 1978.

——. "Galileo and the Tyranny of Truth." In: *The Galileo Affair*, 155–166.

Fölsing, Albrecht. *Galileo Galilei — Prozess ohne Ende: Eine Biographie*. Munich-Zürich, 1983.

Freiesleben, Hans-Christian. *Der Prozess gegen Galilei*, Veröffentlichungen der Gesellschaft Hamburger Juristen 6. Hamburg, n.d. [after 1965].

——. *Galilei als Forscher*. Darmstadt, 1968.

Favaro, Antonio, ed. *Le opere di G. Galilei*, vols. 1–20. Florence, 1890–1909.

Galileo Galilei, 350 anni di storia (1688–1983): Studi e ricerche. Edited by P. Poupard. Rome, 1984.

Galileo Reappraised. Edited by Golini, C. L. Berkeley, 1966.

The Galileo Affair: A Meeting of Faith and Science, proceedings of the Cracow Conference 24–27 May 1984. Edited by G. V. Coyne, M. Heller, and J. Zycinski. Vatican City, 1985.

Graffi, D. "La questione Galileo-Copernicana dopo tre secoli." In: *Accademia Patavina*, 167–175.

Grisar, Hartmann. *Galileistudien: Historisch-theologische Untersuchungen über die Urtheile der römischen Congregationen im Galilei-Process.* Regensburg-New York-Cincinnati, 1882.

Hampe, Johann Christoph. *Die Autorität der Freiheit — Gegenwart des Konzils und Zukunft der Kirche im ökumenischen Disput* III. Munich, 1967.

Hemleben, Johannes. *Galileo Galilei in Selbstzeugnissen und Bilddokumenten.* Hamburg, 1979.

Hurter, Hugo. *Nomenclator Literarius Theologiae Catholicae: Theologos exhibens aetate, natione, disciplinis distinctos* V/1. Oeniponte, 1911.

Jacqueline, Bernard. "La Chiesa e Galileo nel secolo dell'Illuminismo." In: *Galileo Galilei, 350 anni di storia,* 181–195.

Jünger, Ernst. *Strahlungen* II, Sämtliche Werke: Tagebücher III. Stuttgart, 1979.

Koestler, Arthur. *Die Nachtwandler: Das Bild des Universums im Wandel der Zeit.* Bern-Stuttgart-Vienna, 1959.

Kraus, Andreas. *Die naturwissenschaftliche Forschung an der Bayerischen Akademie der Wissenschafter im Zeitalter der Aufklärung,* Bayerische Akademie der Wissenschaften: Philosophisch-hitorische Klasse: Abhandlungen 82. Munich, 1978.

Lalande, Joseph Jérôme. *Voyage d'un françois en Italie fait dans les années 1765/66* V. Venice-Paris, 1769.

——. *Astronomie,* I. Paris, 1771.

Langford, Jerome J. *Galileo, Science and the Church.* Ann Arbor, 1971.

Levi-Donati, G. R. "Il 'caso Galileo' e l'ultimo inquisitore: Giovanni Muzzarelli da Fanano." In: *Memorie scientifiche, giuridiche, letterarie: Accademia nazionale di scienze, lettere e arti di Modena,* series 8, vol. 8 (2005).

Loretz, Oswald. *Galilei und der Irrtum der Inquisition: Naturwissenschaft — Wahrheit der Bibel — Kirche.* Kevelaer, 1966.

Maffei, Paolo. *Giuseppe Settele, il suo diario e la questione Galileiana.* Foligno, 1987.

Metzler, Joseph. "Francesco Ingoli, der erste Sekretär der Kongregation." In: *S. C. de Propaganda Fide Memoria verum 1622–1972.* Rome-Freiburg-Vienna, 1971, 197–232.

Miscellanea Galileiana I-III, Pontificiae Academiae Scientiarum Scripta Varia 27. Vatican City, 1964.

Müller, Adolf. *Der Galilei-Prozess.* Freiburg im Breisgau, 1909.

——. *Galileo Galilei und das kopernikanische Weltsystem.* Freiburg im Breisgau, 1909.

Noack, Friedrich. *Das Deutschtum in Rom seit dem Ausgang des Mittelalters* II. Stuttgart, 1927; new edition Aalen, 1974.

Novità celesti e crisi del sapere: Atti del Convegno Internazionale di Studi Galileiani. Edited by P. Galluzzi. Supplemento agli atti dell'Istituto e Museo di Storia della Scienza Anno 1983, fasc. 2. Florence, 1983.

Di Copernico e di Galileo. Scritto postumo del Fr. Maurizio Benedetto Olivieri, Ex-generale dei Domenicani e Commissario della S. Romana ed Universale Inquisizione, ora per la prima volta messo in luce sull'autografo per la cura di un religioso dello stesso istituto. Bologna, 1872.

Paschini, Pio. "Vita e Opere di Galileo Galilei." In: *Miscellanea Galileiana* I-II, 1–724.

Pastor, Ludwig von. *Geschichte der Päpste seit dem Ausgang des Mittelalters* 13/II. Freiburg im Breisgau, 1929.

Pedersen, Olaf. "Galileo and the Council of Trent: The Galileo Affair revisited." In: *Journal for the History of Astronomy* 14 (1983): 1–29.

Phillips, Georg. *Kirchenrecht* VI. Regensburg, 1864.

Piolanti, Antonio. *L'Accademia di Religione Cattolica: Profilo della sua storia e del suo Tomismo*, Biblioteca per la storia del Tomismo 9. Vatican City, 1977.

Quarto centenario della Nascità di Galileo Galilei, Pubblicazioni dell'Università Cattolica del Sacro Cuori: Contributi III/8. Milan, 1966.

Reusch, Franz Heinrich. *Der Index der verbotenen Bücher: Ein Beietrag zur Kirchen- und Literaturgeschichte* I, II/1 and II/2. Bonn, 1883, 1885, and 1885.

Riccioli, Giovanni Battista. *Almagestum novum...* II. Bonn, 1651.

Ritzler, R. and Sefrin, P. *Hierarchia catholica medii et recentioris aevi* VI and VII. Padua 1958 and 1968.

Saggi su Galileo Galilei. Edited by the Comitato Nazionale per le manifestazioni celebrative del IV. Centenario della nascita di Galileo Galilei. Florence, 1967.

Sarasohn, Lisa T. "French Reaction to the Condemnation of Galileo, 1632–1642." In: *Catholic Historical Review* 74 (1988): 34–54.

Scherz, Gustav. "Niels Stensen und Galileo Galilei." In: *Saggi su Galileo Galilei*, 3–65.

Schmidlin, Josef. *Papstgeschichte der neuesten Zeit* I. Munich, 1933.

Schnürer, Gustav. *Katholische Kirche und Kultur in der Barockzeit*. Paderborn, 1937.

Schuster, Leopold. *Johann Kepler und die grossen kirchlichen Streitfragen seiner Zeit*. Graz, 1888.

Schwedt, Hermann H. *Das römische Urteil über Georg Hermes (1775–1831): Ein Beitrag zur Geschichte der Inquisition im 19. Jahrhundert*, Römische Quartalschrift: Supplementheft 37. Rome-Freiburg-Vienna, 1980.

La scuola Galileiana: prospettive di ricerca, Atti del Convegno di studio di Santa Margherita Ligure (26–28 ottobre 1978). Edited by G. Arrighi et al., Pubblicazioni del Centro di Studi del Pensiero Filosofico del

Cinquecento e del Seicento in Relazione ai Problemi della Scienza 1/14. Florence, 1979.

Soccorsi, Filippo. "Il Processo di Galileo." In: *Miscellanea Galileiana* III, 849–929.

Šolle, Zdenko. *Neue Gesichtspunkte zum Galilei-Prozess,* Österreichische Akademie der Wissenschaften: Philosophisch-Historische Klasse: Sitzungsberichte 361. Vienna, 1980.

Soppelsa, M. *Genesi del metodo Galileiano e tramonto dell'Aristotelismo nella scuola di Padova,* Saggi e testi 13. Padua, 1974.

Spini, Giorgio. "The Rationale of Galileo's Religiousness." In: *Galileo Reappraised,* 44–66.

Taurisano, I. *Hierarchia Ordinis Praedicatorum* I. Rome, 1916.

The Church and Galileo. Edited by E. McMullin. Notre Dame, IN, 2005.

Thiersch, Friedrich von. "Rede über die Grenzscheide der Wissenschaften." In: *Gelehrte Anzeigen der Königlich Bayerischen Akademie der Wissenschaften* 41 (1855): 177–211.

Thomas Aquinas. *Opera Omnia, editio iussu impensaque Leonis XIII. P. M.* Rome, 1882ff.

——. *Opera omnia* III. Edited by R. Busa. Stuttgart-Bad Cannstatt, 1980.

Tiraboschi, Girolamo. *Storia della Letteratura Italiana* X. Rome, 1797.

Tre secoli di storia dell'Arcadia. Rome, 1991.

Vasoli, Cesare. "Sulle fratture del Galileismo nel mondo della controriforma." In: *La scuola Galileiana,* 203–213.

——. " 'Tradizione' e 'Nuova Scienza.' Note alle lettere a Cristina di Lorena ed al P. Castelli." In: *Novità celesti,* 73–94.

Vernacchia-Galli, Jole. *L'Archiginnasio Romano secondo il diario del prof. Giuseppe Settele (1810-1836),* Studi e Fonti per la storia dell'Università di Roma 2. Rome, 1984.

Wallace, William A. "Galileo e i professori del Collegio Romano alla fine del secolo XVI." In: *Galileo Galilei, 350 anni di storia,* 76–97.

Weiland, Albrecht. *Der Campo Santo Teutonico in Rom und seine Grabdenkmäler,* Der Campo Santo Teutonico I. Rome-Freiburg-Vienna, 1988.

Abbreviations of Reference Works Consulted

ADB: *Allgemeine deutsche Biografie*

CSEL: *Corpus scriptorum ecclesiasticorum Latinorum* 1ff. Vienna, 1866ff.

DBI: *Dizionario biografico degli Italiani* 1ff. Rome, 1960ff.

DHGE: *Dictionnaire d'histoire et de géographie ecclésiastiques* 1ff. Paris, 1912ff.

DThC: *Dictionnaire de théologie catholique* I–XV. Paris, 1903–1950.

EC: *Enciclopedia Cattolica* I–XII. Vatican City, 1949–1954.

EncIt: *Enciclopedia Italiana di scienze, lettere ed arti* 1–36. Rome, 1929–1939.

ERSCH/GRUBER: J. S. Ersch and J. G. Gruber, eds. *Allgemeine Enzyklopädie der Wissenschaften und Künste* I/1–99. Leipzig, 1818–1882.

KL2: H. J. Wetzler and H. Welte, eds. *Kirchenlexikon oder Encyklopädie der katholischen Theologie und ihrer Hilfswissenschaften* I–XII. Freiburg im Breisgau, 1882–1903.

LMA: *Lexikon des Mittelalters* Iff. Munich-Zürich, 1980ff.

LThK2: J. Höfer and K. Rahner, eds. *Lexikon für Theologie und Kirche* I–X. Freiburg im Breisgau, 1957–1965.

MICHAUD: L.-G. Michaud, *Biographie universelle ancienne et moderne* 1–45. Paris, no date [1843–1865].

MORONI: G. Moroni, *Dizionario di erudizione storico-ecclesiastica* 1–103. Venice, 1840–1861.

NDB: *Neue deutsche Biographie* Iff. Berlin, 1953ff.

PG: J.-P. Migne, ed. *Patrologiae cursus completus: Series Graeca* 1–167. Paris, 1857–1866.

SOMMERVOGEL: C. Sommervogel, *Bibliothèque de la Compagnie de Jésus* I–IX. Brussels-Paris, 1890–1900; reprint Héverlé-Louvain, 1960.

WURZBACH: C. v. Wurzbach, *Biographisches Lexikon des Kaisertums Oesterreich* 1–60. Vienna 1856–1891.

ZEDLER: J. H. Zedler, ed. *Grosses vollständiges Universal-Lexikon aller Wissenschaften und Künste* 1–64. Halle-Leipzig, 1732–1750.

Bibliographical Supplement

D'Addio, Mario. *Il caso Galilei: Processo — Scienza — Verità.* Rome, 1993, 237.

Baldini, U. "Sul contesto storico e scientifico del caso Settele in margine a W. Brandmüller/E. J. Greipl, *Copernico, Galilei e la Chiesa*." *ComPAcSc* 34 (1996): 21–58.

Battistini, A. *Galileo e i gesuiti: Miti letterari e retorica della scienza.* Milan, 2000, vi + 420.

Beretta, F. *Galilée devant le tribunal de l'Inquisition: Une relecture des sources.* Fribourg, 1998, vii + 320.

———. "Le procès de Galilée et les archives du Saint-Office: Aspects judiciaires et théologiques d'une condamnation célèbre." *RScPhTh* 83 (1999): 441–490.

———, ed. *Galilée en procès, Galilée réhabilité?* Saint-Maurice, 2005, 173.

———. "Melchior Inchofer et l'hérésie de Galilée: censure doctrinale et hiérarchie intellectuelle." In: *Journal of Modern European History* 3 (2005): 23–49.

Bucciantini, M., Camerota, M., and Giudice, F., eds. *Il caso Galileo: una rilettura storica, filosofica, teologica,* Atti del convegno internazionale di studi (Firenze 26–30 maggio 2009). Florence, 2011, xiv + 522.

Christoph Clavius e l'attività scientifica dei gesuiti nell'età di Galileo, Atti del Convegno internazionale, Chieti, 28–30 aprile 1993. Edited by U. Baldini. Rome, 1995, 318.

Cioni, M. *I documenti galileiani del S. Uffizio di Firenze*. Florence, 1996, xxviii + 80.

Contro Galileo, Alle origini dell'"affaire." Florence, 1995, 220.

Copernico e la questione copernicana in Italia dal XVI al XIX secolo. Edited by L. Pepe. Florence, 1996, xiv + 294.

Damanti, A. *Libertas philosophandi: Teologia e filosofia nella Lettera alla granduchessa Cristina di Lorena di Galileo Galilei*. Rome, 2010, 562.

l'documenti vaticani del processo di Galileo Galilei (1611–1741). New expanded edition, revised and annotated by S. Pagano. Vatican City, 2009, cclvii + 332.

Donati, A. *Le motivazioni teologiche della condanna di Galileo Galilei*. Perugia, 2010, 168.

Fantoli, A. *Galileo for Copernicanism and for the Church*. Vatican City, 1994, xix + 540.

——. *Galileo e la chiesa: una controversia ancora aperta*. Rome, 2010, 255.

Feldhay, R. *Galileo and the Church: Political Inquisition or Critical Dialogue?* Cambridge, 1995, viii + 303.

Finocchiaro, M. A. *Retrying Galileo, 1633–1992*. Berkeley, 2005, 498.

Fischer, K. *Galileo Galilei: Biografie seines Denkens*. Stuttgart, 2015, 279.

Frajese, V. "A proposito del processo a Galileo; Il problema del precetto Seghizzi." In: *Annali della Scuola Normale Superiore di Pisa, Classe di Lettere e Filosofia, S. 5* vol. 1 (2009): 507–533.

——, ed. *Il processo a Galileo Galilei: Il falso e la sua prova*. Brescia, 2010, 120.

Hedley Brooke, J. *Science and Religion: Some Historical Perspectives*. Cambridge, 1993, x + 422.

Il processo a Galileo Galilei e la questione Galileiana: A vent'anni dalla morte di Luigi Firpo, Atti del convegno internazionale, 26–27 marzo 2009. G. M. Bravo and V. Ferrone, eds. Rome, 2010, xvii + 318. (Contents: G. M. Bravo, "Premessa: Galileo, la rivoluzione scientifica, la libertà di ricerca e Luigi Firpo"; A. Ferrari, "Galileo e il cannocchiale"; P. Galluzzi, "Il 'caso Galileo' "; A. Prosperi, "L'Inquisizione e Galilei"; W. Shea, "Galileo e la Chiesa romana"; F. Beretta, "Il processo di Galileo: Due nuove edizioni dei documenti"; F. Motta, "Epistemologie cardinalizie: Ipotesi, verità, apologia"; F. Raimondi, "Scienza politica in Galilei: Critica dell'eresia e nascita dell''ideologia' "; M. Camerota, "Galileo, la scienza e la teologia"; M. Matteoli, "Giordano Bruno e la geometria dell'infinita materia"; G. Ernst, "Galileo, Campanella e le dottrine celesti"; L. Guerrini, "Raffaello delle Colombe et les origines de la polémique anti-galiléenne à Florence (1610–1615)"; M. Simonazzi, "Elementi di retorica galileiana nell'opera di Mandeville?"; V. Ferrone, "Lilluminismo e il caso Galileo: Breve storia di un problema mal posto?"; S. Suppa, "Galileo Galilei nell'*Encyclopédie*"; A. Fantoli, "Il caso Galileo: Una questione chiusa?"; A. Zambarbieri, "Libertà della ricerca e divieti ecclesiastici: Una rivisitazione modernista del caso Galilei"; M. Bucciantini, "Storie d'Italia: La riconquista di Galileo"; G. Lolli, "La retorica di Galileo"; D. Ronchi della Rocca, "La questione galileiana, oggi.")

Intrieri, L. "La vicenda di Galileo nel suo tempo." *Vivarium* 1 (1993): 533–553.

La Dous, L. *Galileo Galilei: zur Geschichte eines Falles.* Regensburg, 2007, 173.

Lamanna, M. and P. Ponzio. "Benet Perera e Galileo Galilei sull'unità del vero." In: *Freiburger Zeitschrift für Philosophie und Theologie* 64 (2017): 301–322.

Levi-Donati, G. R. "Il 'caso Galileo' e l'ultimo inquisitore: Giovanni Muzzarelli da Fanano." In: *Memorie scientifiche, giuridiche, letterarie,* Accademia nazionale di scienze, lettere e arti di Modena, Series 8, vol. 8 (2005), 5.28.

Martinez, R. "Il significato epistemologico del caso Galileo: due diverse concezioni della scienza." In: *ActPh* 3 (1994): 45–74.

Mayaud, P.-N. *Le conflit entre l'astronomie nouvelle et l'écriture sainte aux XVIe et XVIIe siècles: Un moment de l'histoire des idées: Autour de l'affaire Galilée.* Paris, 2005, 6 vols., in 5 (440, 404, 1330, 259, 409), bibliography, indices.

Mayer, T.F. "The Status of the Inquisition's Precept to Galileo (1616) in Historical Perspective." In: *Nuncius* 24 (2009): 61–96.

——. *The Trial of Galileo, 1612–1633*. Toronto, 2012, 224.

Millefiorini, P. "Galileo al di là delle polemiche." In: *RaT* 35 (1994): 497–505.

Moss, J. D. *Novelties in the Heavens: Rhetoric and Science in the Copernican Controversy*. Chicago-London, 1993, xiv + 353.

Pesce, M. "L'indisciplinabilità del metodo e la necessità politica della simulazione e della dissimulazione in Galilei dal 1609 al 1642." In: *Disciplina*, 161–184.

——. *L'ermeneutica biblica di Galileo e le due strade della teologia Cristiana*. Rome, 2005, 240.

Porter, S. K. *Remembering Galileo*. Lanham, MD, 1995, 204 illustrated.

Prosperi, A. "Processi a Galileo." In: *Belfagor* 65 (2010): 161–182.

Redondi, P. *Galileo eretico*. Turin, 1983; reprinted Rome-Bari, 2009, 479.

Riaza Morales, J. M. *La Iglesia en la historia de la ciencia*. Madrid, 1999, xix + 319.

Rosen, E. *Copernicus and His Successors*. London and Rio Grande, 1995, 244.

Salucci, A. "La metafora del libro della natura in Galileo Galilei." In: *Angelicum* 83 (2006): 327–375.

Shea, W. R. and Artigas, M. *Galileo Galilei: Aufstieg und Fall eines Genies*. Darmstadt Wissenschaftliche Buchgesellschaft, 2006, 236.

Sierotowicz, T. "'Un solo dio è quello che la sa tutta,' Esercitazioni retoriche di Galileo Galilei." In: *Asprenas* 58 (2011): 3–4, 345–362.

Simoncelli, P. *"Storia di una censura: "Vita di Galileo" e concilio Vaticano II*. Milan, 1994, 153.

Smith, G. *Galileo: A Dramatised Life*. London, 294.

Šolle, Zdenko. "Galileo Galilei und die Länder nördlich der Alpen," compiled by K. Ferrari D'Occhieppo, edited by G. Hamann and H. Grössing. In: *AnzÖAk* 131 (1994): 191–227.

Speller, J. *Galileo's Inquisition Trial Revisited*. Frankfurt, etc.: Peter Lang, 2008.

Tanzella-Niti, G. "Giovanni Paolo II e Galileo Galilei." In: *Annales Theologicae* 24 (2010): 2, 411–424.

The Church and Galileo. Edited by E. McMullan. Notre Dame, IN, 2005, xii + 391.

Bibliography of Galileo's Works

according to the Edizione Nazionale

1. *Theoremata Circa Centrum Gravitatis Solidorum* (Reflections on the center of gravity of solid bodies), 1586. Partially published in no. 38, first complete publication by Favaro/Lungo.

2. *Juvenilia*, ca. 1586-1589. First published by Favaro/Lungo.

3. *La Bilancetta* (The scale), 1586. First published by Favaro/Lungo.

4. *Tavola delle Proporzioni delle Gravita in Specie dei Metalli e delle Gioie, pesate in Aria ed in Acqua* (Table of specific gravities of metals and jewels), 1586. First published by Favaro/Lungo.

5. *Postille ai libri "De Sphaera et Cylindro" di Archimede* (Remarks on Archimedes's *On the Sphere and the Cylinder*), ca. 1586. First published by Favaro/Lungo.

6. *Contro il portar la Toga* (Against wearing the toga), ca. 1589–1592. First published in: M. F. Berni, *Opere burlesche* III (Florence, 1723), 177–187.

7. *De Motu* (On movement), 1590. First complete publication by Favaro/Lungo.

8. *Due Lezioni all'Accademia Fiorentina circa la figura, sito e grandezza dell'Inferno di Dante* [Two lectures to the Academy of Florence about the form, location, and size of Hell in Dante); exact time of

composition uncertain, at any rate before December 1592. First published by O. Gigli. Florence: F. Le Monier, 1855.

9. *Sonetto* (Sonnets), date uncertain. First complete edition by Favaro/Lungo according to the extant manuscripts and several seventeenth-century printings.

10. *Argomento e traccia d'una comedia* (Plot and sketch for a comedy), 1592–1599. First published by Favaro/Lungo.

11. *Breve istruzione all'architettura militare* (Short Instruction about military architecture), 1592–1593. First published by Favaro/Lungo.

12. *Trattati di architettura militare e di fortificazioni.* (Treatise on fortresses), 1592–1593. First published in: G. B. Venturi, *Memorie e lettere inedite finora o disperse di Galileo Galilei* I. Modena: G. Vincenzi, 1818.

13. *Le Mechaniche* (Mechanics), 1597–1598. First published by Favaro/Lungo.

14. *Lettera a Iacopo Mazzoni* (Letter to J. Mazzoni), May 30, 1597. First published by Favaro/Lungo.

15. *Postille all'Ariosto* (Remarks on Ariosto), ca. 1603. First published by Favaro/Lungo.

16. *Trattato della Sfera Ovvero Cosmografia* (Treatise about the heavenly sphere or cosmography), 1604–1605. First published by B. Sari, Rome: D. Grialdi (publisher) and N. A. Tinazzi (printer), 1656.

17. *De Motu Accelerato* (On acceleration), 1604. First complete publication by Favaro/Lungo. Partial publication in no. 38.

18. *Frammenti di Lezioni e di Studi sulla nuova Stella dell'Ottobre 1604* (Fragments of lectures and studies about the new star in October 1604), 1604. First published in: G. B. Venturi, II. Modena: Vincenzi, 1821.

19. *Le Operazioni del Compasso Geometrico e Militare* (The operations of the geometrical-military compass). Padua: P. Marinelli, 1606.

20. *Difesa di Galilei Nobile Fiorentino Contro alle Calunnie & Imposture di Baldessar Capra Milanese* (Defense ... against the calumnies and deceptions of Baldassar Capra of Milan). Venice: T. Baglioni, 1607.

21. *Sidereus Nuncius* (Starry messenger). Venice: T. Baglioni, 1610.

22. *Le Matematiche nell'Arte Militare* (Mathematics in military science), 1610. First published by Favaro/Lungo, authenticity not absolutely certain.

23. *Considerazioni al Tasso* (Reflections on Tasso), before 1614. First published by P. Pasqualoni, Rome: Pagliarini, 1794. Authenticity probable according to Favaro/Lungo.

24. *Pianeti Medicei* (The Medici planets), 1610. First published by Favaro/Lungo.

25. *Theorica Speculi Concavi Sphaerici* (Teaching about the spherical concave mirror), ca. 1610. First published by Favaro/Lungo.

26. *Analecta Astronomica* (On astronomical topics), ca. 1612. First published by Favaro/Lungo.

27. *Delle Cose che stanno in su l'Acqua o che si muovono* (About objects that stand on water or move). Florence: C. Giunti, 1612.

28. *Istoria e Dimostrazioni intorno alle Macchie Solari e Loro Accidenti Comprese in tre Lettere Scritte all'Illustrissimo Signor Marco Velseri Linceo Duumviro d'Augusta* (Three letters about sunspots to Marcus Welser). Rome: G. Mascardi, 1613.

29. *Proposta per la determinazione delle Longitudini* (Suggestions for determining longitudes), 1612. First published by Favaro/Lungo.

30. *Lettera a Tolomeo Nozzolini* (Letter to T. Nozzolini), January 1613. First published by Favaro/Lungo.

31. *Lettera a D. Benedetto Castelli* (Letter to Don B. Castelli), December 21, 1613. First partial publication in: G. Targioni Tozzetti, *Notizie degli aggrandimenti delle scienze fisiche* I/1 (Florence, 1780), 22–26. First complete publication by Favaro/Lungo.

32. *Lettere a Mons. Piero Dini* (Letters to Msgr. P. Dini), February 16 and March 23, 1613. First published in: I Morelli, *I codici manoscritti volgari della Libreria Naniana* (Venice, 1776), 191–194, 195–201.

33. *Lettera a Madama Cristina di Lorena Granduchessa di Toscana* (Letter to Christine of Lorraine, Grand Duchess of Tuscany), 1616. First published in a Latin translation in Strasbourg: Elzevier, 1636.

34. *Discorso del Flusso e Riflusso del Mare* (On the tides), 1616. First published by Favaro/Lungo.

35. *Discorso delle Comete* (On the comets). Published under the name of Galileo's student M. Guiducci. Florence: Cecconcelli, 1619.

36. *Il Saggiatore* (The goldsmith's scale). Rome: G. Mascardi, 1623.

37. *Dialogho sopra i due massimi sistemi del mondo* (Dialogue about the two major cosmologies). Florence: G. B. Landini, 1632.

38. *Discorsi e dimostrazioni matematiche intorno a due nuove scienze* (Discourses and mathematical proofs concerning two new branches of natural science). Leyden: Elzevier, 1638.

39. *Le operazioni astronomiche* (Astronomical calculations), 1638. First published by Favaro/Lungo.

Chronological Table

1564	Conclusion of the Council of Trent — Galileo born on February 15 in Pisa — Calvin and Michelangelo died.
1569	Cosimo I, Grand Duke of Tuscany (†1574).
1571	Johannes Kepler born (†1630).
1572	Gregory XIII, Pope (†1585).
1574	Galileo family resettled in Florence — Francesco Maria Grand Duke of Tuscany (†1587).
1575	Galileo a student at the monastery school in Vallombrosa (until 1578).
1581	Galileo a student in Pisa (until 1585).
1585	Sixtus V, Pope (†1590).
1587	Galileo's first trip to Rome, acquaintance with Clavius — Ferdinando I, Grand Duke of Tuscany (†1609).
1589	Galileo's appointment as professor in Pisa.
1590	Election of Popes Urban VII and Gregory XIV (†1591).
1591	Galileo's Father Vincenzo died, he himself moved to Florence. — Innocent IX, Pope (†1592).
1592	Galileo's appointment as professor in Padua — Clement VIII, Pope (†1605).
1596	René Descartes born (†1650).
1600/1601	Birth of Galileo's daughters Virginia and Livia — Tycho Brahe died (*1546).
1605	Leo XI and Paul V (†1621), Popes.
1606	Birth of Galileo's son Vincenzo.
1609	Construction of Galileo's telescope, first observations of the heavens — Cosimo II, Grand Duke of Tuscany (†1621).

1610	Appointment as First Mathematician and Philosopher of the Grand Duke of Tuscany — Moved to Florence.
1611	Galileo's second trip to Rome.
1614	Repeated attacks on Galileo from the pulpit by the Dominican Caccini.
1616	Galileo's third trip to Rome — Condemnation of Copernicus and prohibition by the Inquisition against teaching heliocentrism.
1617	Construction of a binocular telescope by Galileo for determining distances at sea.
1618	Appearance of three comets — Outbreak of the Thirty Years' War.
1621	Ferdinand II, Grand Duke of Tuscany (†1670), until 1627 tutelary rule by the Grand Duchess Mother Christina — Gregory XV, Pope (†1623) — Cardinal Bellarmine died (*1542).
1623	Urban VIII, Pope (†1643) — Galileo's fourth trip to Rome — Beginning of work on the *Dialogo* — Blaise Pascal born (†1662).
1630	Intervention of Gustav Adolf of Sweden into the Thirty Years' War.
1632	Victory of Gustav Adolf over Tilly at Rain am Lech; Tilly fatally wounded — Death of Gustav Adolf in the Battle of Lützen — Galileo summoned before the Inquisition.
1633	Trial against Galileo, his condemnation and recantation in Rome — Sojourn in Siena, return to Arcetri.
1634	Death of Galileo's favorite daughter, Virginia (Sister Maria Celeste), in Arcetri.
1635	Entrance of France into the Thirty Years' War as an opponent of the Habsburgs.
1636	Thomas Hobbes visited Galileo.
1637	Galileo goes blind.
1638	Visit of John Milton to Galileo.
1639	Vincenzo Viviani becomes Galileo's fellow lodger.
1641	Evangelista Torricelli (†1647) also becomes Galileo's fellow lodger.
1642	Galileo died on January 8.

Index

About the Author

Cardinal Walter Brandmüller (b. 1929), former president of the Pontifical Commission for Historical Sciences, was born in the Federal Republic of Germany. He was ordained a priest on July 26, 1953, and earned a doctorate in theology in 1963. In 1971, he taught medieval and modern Church history at the University of Augusburg and continued until 1997, during which period he served as parish priest in Walleshausen. In June 1998, he was appointed president of the Pontifical Commission for Historical Sciences, a post he held until he retired in December 2009. In 2001, he was elected president of the International Commission for Comparative Ecclesiastical History. He is a world-renowned scholar of Church history and has published many books and articles on the Crusades, the Spanish Inquisition, and the Reformation. In November 2010, he was ordained a bishop and appointed titular archbishop of Caesarea in Mauretania. He was created and proclaimed a cardinal by Pope Benedict XVI in the consistory of November 2, 2010.

Sophia Institute

Sophia Institute is a nonprofit institution that seeks to nurture the spiritual, moral, and cultural life of souls and to spread the gospel of Christ in conformity with the authentic teachings of the Roman Catholic Church.

Sophia Institute Press fulfills this mission by offering translations, reprints, and new publications that afford readers a rich source of the enduring wisdom of mankind.

Sophia Institute also operates the popular online resource CatholicExchange.com. *Catholic Exchange* provides world news from a Catholic perspective as well as daily devotionals and articles that will help readers to grow in holiness and live a life consistent with the teachings of the Church.

In 2013, Sophia Institute launched Sophia Institute for Teachers to renew and rebuild Catholic culture through service to Catholic education. With the goal of nurturing the spiritual, moral, and cultural life of souls, and an abiding respect for the role and work of teachers, we strive to provide materials and programs that are at once enlightening to the mind and ennobling to the heart; faithful and complete, as well as useful and practical.

Sophia Institute gratefully recognizes the Solidarity Association for preserving and encouraging the growth of our apostolate over the course of many years. Without their generous and timely support, this book would not be in your hands.

www.SophiaInstitute.com
www.CatholicExchange.com
www.SophiaTeachers.org